Risk-Led Safety

Risk-Led Safety: Evidence-Driven Management, Second Edition

Authored by
Duncan Spencer and Chris Jerman

CRC Press
Taylor & Francis Group
Boca Raton London New York

CRC Press is an imprint of the
Taylor & Francis Group, an **informa** business

CRC Press
Taylor & Francis Group
6000 Broken Sound Parkway NW, Suite 300
Boca Raton, FL 33487-2742

International Standard Book Number-13: 978-0-367-42998-0 (Hardback)
International Standard Book Number-13: 978-0-367-42266-0 (Paperback)

Library of Congress Cataloging-in-Publication Data

Names: Spencer, Duncan (Safety engineer), author. | Jerman, Chris, author.
Title: Risk-led safety : evidence-driven management / by Duncan Spencer, Chris Jerman.
Description: Second edition. | Boca Raton : CRC Press/Taylor & Francis Group, 2020. | Includes bibliographical references and index.
Identifiers: LCCN 2019031436 (print) | LCCN 2019031437 (ebook) | ISBN 9780367422660 (paperback ; acid-free paper) | ISBN 9780367823429 (ebook)
Subjects: LCSH: Industrial safety. | Risk management. | Hazard mitigation.
Classification: LCC T55 .S66 2020 (print) | LCC T55 (ebook) | DDC 658.4/08--dc23
LC record available at https://lccn.loc.gov/2019031436
LC ebook record available at https://lccn.loc.gov/2019031437

Visit the Taylor & Francis Web site at
http://www.taylorandfrancis.com

and the CRC Press Web site at
http://www.crcpress.com

Duncan Spencer

"To my wife Nicki and children Chris and Sophie, a simple thank you for being there. And to my friend Chris, whose constant challenge to rethink things drives my understanding forward to new heights."

Chris Jerman

"Thanks to my wife Claire for believing in me and patiently listening to my rants, to Jim Brough for being both the most inspirational manager and strongest leader I ever met and to Doctor John Herbert for providing me with the best direction in those early years at the EEF. Lastly to Duncan and the Human Applications crew for being on the same wavelength."

Contents

Foreword

Many organisations feel that they have plateaued with regard to the potential improvements from their safety systems and processes with much of modern safety management now focused on the gains to be had from engaging more actively with the workforce and their behaviours. This means addressing the culture, which is perhaps best defined as "the way we do things around here on a typical day". We often say re an organisation's policies, processes and procedures that once the organisation has hit diminishing returns, it's all about objective learning culture and effective transformational leadership.

However, before, or alongside, embarking upon a culture and behavioural journey, it is essential to reflect clearly on just how well your business is actually being served by its safety management systems. The paperwork looks fine and dandy – but how effective are these systems and processes really? How user-friendly are they? How integrated and joined up are the major elements? Are there simple but effective changes and tweaks that can genuinely minimise the number of issues and cultural and behavioural methodologies needing to be addressed?

This excellent, practical, book, by two of Europe's leading thinkers on this subject, contains simple but effective tools to raise a number of key questions – thus allowing the reader to judge their own company's processes and to identify any shortcomings or conflicts. For any reader just embarking on their journey into the world of safety and health – or even for those of us who have been around for a while now – it proves itself a thought-provoking and vital source of information.

I won't embarrass the authors by mentioning just how many years of practical and tried and tested experience they have between them, as it'll become readily apparent to any reader. It's a very welcome and fresh take on a vital subject.

Dr. Tim Marsh, July 2019

Preface

Health & safety does not need to be complicated. Indeed, if one goes back to the Health and Safety at Work etc. Act of 1974, it is rather a clear document and very much fit for purpose. However, in the years since the passage of the Act, many practitioners have inadvertently imposed on themselves excess health & safety bureaucracy due in part to risk aversion, media hype or insurance concerns. In speaking with practitioners in different parts of the country, many have told me that they would welcome more guidance on what they needed to do in terms of health & safety in order to be compliant with the Act. This excellent practical handbook addresses this gap by providing that needed guidance. It should be essential reading for anyone active in the health & safety area, be they practitioners, politicians or other considered stakeholders.

Ragnar E. Löfstedt is Professor of Risk Management at King's College London and Director of King's Centre for Risk Management.

Authors

The authors have over 45 years of experience between them. They have applied and developed these ideas in many different organisations and industries. This is a no-nonsense book by health & safety practitioners who have learnt their craft from doing it.

Duncan Spencer is the Head of Advice & Practice at the Institution of Occupational Safety & Health, the world's leading membership body for safety professionals. His more recent experience includes being responsible for health & safety governance in the John Lewis Partnership and the Safety Manager for the Waitrose Division. This included responsibility for retail, distribution, head office functions, leisure and for the Leckford farming estate.

Duncan's previous experience includes a short military career, being the director of an outdoor adventure centre, a college lecturer and a safety risk consultant. In this capacity, his work was in large multiple site and complex organisations in the High Street retail, distribution, social housing, manufacturing and public service sectors (police, military and health care) at home and abroad.

Duncan regularly writes magazine articles and makes presentations at conferences that challenge people to think again about their approach to safety risk management.

Chris Jerman began his safety career in the foundries of South Yorkshire, having spent his formative years in production management. Schooled largely by the EEF in Sheffield, Chris saw the move from the rule-driven regime of the Factories Act to the more modern risk-led approach offered by the Management Regulations. Where many may have seen risk assessment as a burden, Chris took this new requirement as an opportunity, realising at a very early stage that being risk-centric was the way forward.

Moving from heavy industry, Chris spent several years in consultancy and training, working right across Europe and into the Far East, during which time he had the opportunity to see how various risk models worked, the keys to their success and the common pitfalls too. Joining John Lewis in 2005, Chris had the perfect opening to put into practice all that he had learned, benefiting from all of the support of a renowned and forward-looking business.

After a brief period as Group Head for Safety for the bookmaker Ladbrokes Coral, Chris is now working in partnership with Professor Tim Marsh and Jason Anker MBE in their new venture.

Chris has built his reputation on being forthright, often moving where others fear to tread. Chris' drive is to do things pragmatically and is brave enough to do things differently to achieve it.

Introduction

Duncan Spencer and Chris Jerman

Our aim was to write this book in a style that reads easily and logically. We have not wished it to be a highbrow academic work; instead, we have written it as a general guide for the benefit of managers who are not specialists in health & safety, as well as practitioners who are. We express in our own words what we have found to work in practice and any similarity to the work of others is purely coincidental, with the exception of a few ideas that were given root in our time working with Human Applications Ltd., one of the UK's leading safety risk and ergonomics consultancies. In addition, we have gained experience in many different industries and organisations. Chris comes from a background of heavy industry, manufacturing and production management, whilst Duncan's prior experience comes from military service, sales, sheet metal manufacturing, outdoor adventure pursuits and further education.

We have tried and tested these ideas over many years. Our combined experience includes work for police, fire and rescue services, the National Health Service (NHS), large government agencies, retail, distribution, food manufacturing and many other industries in the UK and abroad. Our work for the same leading UK retailer enabled us to fully integrate the approach and provide new and improved standards of health & safety governance in the company. We implemented a task-based risk-led safety management system across the whole business, including distribution, manufacturing, farming and leisure, as well as the food, clothing and furniture retail arms. We believe the approach presented is a pragmatic interpretation and application of UK health & safety legislation. Our experience of working abroad led us to believe that the principles we describe in this book are equally useable in different countries and fully support organisations to meet local legislative requirements.

Working with the principles and concepts contained in this book, we have come to realise that the crucial element missing from guidance is the "how to". So, often guidance informs organisations what is needed and why but fails to cover how to do it well enough. We hope that this book goes some way to fill that gap. We would not be so bold as to say that our thoughts are the only way to manage safety risk, but they constitute a way that we know works in practice. Some readers may find some of the ideas challenging, but if on reflection if they are left with the belief that what they are presently doing does not need changing, such an affirmation is useful. We aimed to write this book and this second edition in easy-to-understand language, so it can be easily read from cover to cover. However, we would respectfully suggest that some of the key ideas in its pages deserve some time for considered reflection.

It is interesting to note that some organisations we know that have adopted a similar approach to the one described in this book have subsequently won prestigious safety awards. We are also bold enough to claim that these ideas have found favour with senior UK Health & Safety Executive officials and Local Authority Environmental Health Officers. They have provided verbal endorsement that our pragmatic risk-led approach also fulfils UK legal compliance requirements.

We recognise that the size of the organisation is important to consider when deciding how far and possibly how complex a safety management system needs to be, but nonetheless our experience indicates the ideas in this book are just as relevant to small and large organisations alike. We hope that all safety professionals and operational managers will find value from this work. In particular, organisations whose activities have pushed them passed the "tipping point", where simple "check box" safety is no longer sufficient, or complex multiple sited organisations where management control is more centralised and a greater reliance is placed on local managers to do the right thing on behalf of the head office.

We start the book with our view on what has gone awry in the safety culture of the UK since 1974. This illustrates how some myths are created and why many UK practitioners have perhaps inadvertently found themselves "gold plating" their response to their safety issues. We consider how, in our view, there is nervousness in management circles where managers do not feel confident to argue that they have controlled risk well enough, fearful that they might get caught out somehow. This discussion sets the backdrop for the ideas in the rest of the book. We describe a problem-solving approach that is truly evidence based.

In Chapter 2, we start to challenge some of the commonly held views by re-examining some legal principles from UK legislation and how they may be used as "tools" across the globe. We believe that the precise and careful use of language is the key to a successful safety management system, a theme that continues throughout the book.

Chapters 3, 4, 5, 6 and 7 are the core chapters describing the "how to" and deal with identifying risk, the pitfalls to avoid in risk assessment, key elements of a task-based risk assessment, prioritising what you find and deciding what to do. Chapter 3 includes a logical process for defining which risk assessments an organisation should have, but perhaps more cleverly, it also identifies those that it does not need. Chapters 4 and 5 began as one chapter in concept, but we soon discovered that we could not argue what we believe to be good practice (Chapter 5) without first discussing common mistakes made with risk assessment technique (Chapter 4). Chapter 6 argues that risk assessment is merely a tool that enables risk management and when done well can provide opportunity to prioritise and thereby facilitate the targeting of limited resources. In Chapter 7, our discussion about controls is not about types of control but much more about how control choices are made.

We make no apologies if you are left with the impression after reading this book that good safety is really just about employing techniques that facilitate good management. Chapter 8 explains this central belief in more detail. At one national conference we both attended, we saw three powerful speakers argue the probity of different approaches to safety management. One argued a task-led approach, another establishing good practice while conducting accident investigations and the third person promoted ways to assure safe behaviour. It struck us that there was one central common theme in all of these methods. What underpinned the success of any approach to safety is how far managers are engaged by it.

Chapter 9 explains how reactive safety should be second place to risk management and is complementary to a risk-led approach. We also spend some time discussing the problems and significant flaws that come with a reliance on reactive incident data and argue that the focus should be on reporting leading indicators to the board. In Chapter 10, we suggest why you cannot report leading indicators unless you have a management system that is truly risk driven. Chapter 10 includes a discussion of the merits of obtaining external certification and considers the wider issues of managing risk after it has been found. It also offers fresh ideas about the agenda for health & safety committees and the responsibility of boards.

In the final chapter, we recognise that having promoted our ideas for building a risk-driven safety management system that drives evidence-based decision-making, we cannot finish the book without providing some advice on how a safety professional can make it a reality in their organisation. We have pooled our collective experience as consultants and corporate safety managers to provide basic advice on how the desire for change can be created and how a changed project can be managed effectively. We discuss simple actions and how to overcome the typical obstacles. Of course, project management, making things happen, is an important and complex subject in its own right and demands a lifelong endeavour for perfection.

You will note that the focus in this book is on safety rather than occupational health, on physically harmful hazards rather than those that are physiologically or psychologically damaging. This is an approach that was born out of necessity to ensure clarity. We certainly believe that the ideas are just as applicable to occupational health risk issues. On occasions you will note that we see a difference between task-orientated risk-led safety that we promote in this book and conditions safety (i.e. control of physical hazards). That is because we believe that conditions safety is about fixing the obvious physical hazards in the working environment while risk assessment is about finding out whether more needs to be done to make tasks safe or to control employee exposure to physiological or psychological hazards. Managing safely is what we do to achieve these goals. We would not like the reader to be left with the impression that a task-based risk-led approach is all that is needed; there will always be a need for the safety tour to identify holes in floors, loose handles and broken guards and the like too.

So what will you get from reading on? This book champions an approach that has worked in many industries and can work for you too. If you adopt it, then you will enjoy the benefits of a lean system of management that focuses on the most significant safety issues. You will learn about a system that cuts out needless bureaucracy and leaves you with what you actually need. In the UK, it will empower you and your organisation to confidently apply the principal requirements of the Health & Safety at Work etc. Act 1974 in a way that much reduces your reliance on external advice and guidance. In other countries, its principles will help you to meet local legal requirements. It is a system that produces evidence as to what matters and what does not. Those we have known who have engaged with these ideas have found that not only did it save them money and effort, to their delight, it also further improved their safety performance.

One last point to note: the content of this book does not describe a system that is intended to compete with a national or an international standard like ISO 45001 – Occupational Health & Safety. But it does describe an approach that can be successfully employed to support compliance with standards like these.

We cannot finish this introduction without a note of thanks to our ex-colleagues and friends at Human Applications Ltd. We are grateful for the significant part they played in our early personal development and for providing the foundations for the ideas contained in this book. In particular, our thanks go to Bernie Catterall and Nigel Heaton.

1

It's easy to be misled – lessons from the United Kingdom and Europe

In this chapter, we will explore some of the lessons from UK history and the application of the European approach to occupational health & safety. It is the direct experience of the authors that these considerations provide learning that has great value in other countries despite differences in legislative requirements. The UK experience provides a perspective on the flaws that may arise in applying a pure compliance-driven approach to safety. It provides the foundation for the risk management concepts discussed throughout this book.

The authors propose that when prescriptive legislation was replaced in the UK in 1974, many found it uncomfortable. They sought prescriptive direction from other sources. This chapter makes the case for health & safety professionals to learn how to confidently reduce their reliance on external guidance and build confidence in defining and managing their business's unique health & safety risk profile.

The UK is considered by some to be the world leader in health & safety. While UK law and culture may be more developed or mature than in some other countries, they are not flawless. In an effort to protect people and drive up standards, sometimes UK organisations become overzealous in their interpretation of what is required. While this chapter is UK-centric, it describes issues that other countries ought to be aware of and may even share to some degree.

How did we get to where we are?

In the UK, prior to the Health & Safety at Work etc. Act 1974 (HSWA74), safety legislation was prescriptive in nature. It was Parliament that agreed upon and set the rules for the workplace, leading to a huge number of reactionary regulations being placed on the statute book. Not only did the large number of regulations cause confusion and frustration, but some pieces of legislation also contradicted the requirements of others. In the late 1950s, pressure from business resulted in Parliament trying to trim down safety law into the Factories Act 1961 and the Offices, Shops and Railway Premises Act 1963.

These Acts still relied upon Parliament writing prescriptive legislation to impose safety requirements in workplaces. It simply didn't work. The law was written in terms of "you will not" or "you must". Some employers were found to exploit the obvious loophole: if there wasn't a specific law saying it can't be done, then by exception it could. Although these Acts were a welcome simplification, they did not significantly reduce the numbers of workplace fatalities. The pace of technological change and the drive to find new competitive working practices outpaced Parliament's capacity to legislate. The diversification and specialisation of industry meant that politicians could not sufficiently understand the changing risk profile well enough to produce "good" law. Additionally, it was demonstrated that the enforcement of the law was inconsistent across the country. Something had to be done.

The Robens Committee talked sense

It was in this backdrop that between 1970 and 1972, The Robens Committee[1] reviewed the standards of safety and occupational health control in the UK. They examined what was wrong with prescriptive legislation and investigated whether another approach would be more fruitful in controlling serious safety issues in business. The conclusion was that there was too much law, and attempts to cover every contingency had led to unnecessary elaboration and too much detail and complexity. The review drew the conclusion that a new legal approach was necessary. Many would argue that these findings have strong echoes even today.

Lord Robens came to the conclusion that the only people who truly understood the safety and occupational health issues at their workplaces were the people who managed and worked in them. Some of the reasons for it are listed in Table 1.1.

The Robens report recommended a new way, suggesting that it should be the employer who must predict what could go wrong and implement preventative measures. This is the central philosophy that underpins much of the content of HSWA74. In other words, his recommendation transferred the primary responsibility of making safety rules from Parliament to the decision-makers within each individual organisation. Since the implementation of HSWA74, it has been the legal responsibility of managers and workers to identify what incidents may occur in their organisation and implement their own rules to prevent it.

In order to comply with HSWA74, managers have to interpret how the general requirements of the Act and its "daughter" Regulations (see below) apply to their operational circumstances. This demands competency and confidence; however, between 1974 and 1992 when the first European-led regulations were added to the statute, many businesses struggled to understand what was expected of them. Anecdotally, they looked to others for inspiration and watched criminal cases and considered any relevant civil case precedents to indicate how far they should go.

Unfortunately, for many businesses, perhaps predominantly smaller ones that could not afford a full-time health & safety professional, this change left them uncomfortable. The demise of prescriptive law left them feeling exposed, and they lacked the understanding of how to comply with the new legal framework. This lack of knowledge undermined confidence and led to a

TABLE 1.1

Parliament- versus employer-led safety

Parliament-led safety rules	Employer-led safety rules
Significantly, time lagged between recognising an issue and legislating to control it due to parliamentary process: raising a white paper, debating and amendment by consensus and ratification. This often diluted the effect of the legislation.	Controls can by implemented very quickly and proactively once the issue is identified and in the context of the local requirements of the operational circumstances.
Cannot hope to keep pace with new working methods and technological change. Statute will always lag behind technological advances and changes in the workplace.	When incorporated into business planning, preventative measures can be designed before new technology or processes are started.
Most members of Parliament have not worked in the industry but they are legislating, which diminishes understanding and contextual knowledge.	Managers and their workers do understand local conditions and circumstances and can make their rules more specific and effective.
Tendency for legislation to be written after emotive pressure from the public that may reflect what is reported in the popular press rather than what the true facts are.	Those who should know the true facts write safety rules.

legacy that has been present to some degree in the UK ever since. Perhaps this has led to the demands for the Health and Safety Executive (HSE) to provide detailed guidance and for industry groups to produce standards or "risk-solving products" from businesses eager to exploit the vacuum. This lack of confidence and understanding has not been helped by inadequate teaching of this point to people starting their career in the health & safety profession.

Europe gets in on the act

Moving on in the timeline of UK health & safety legal history, the formation of the European Union from the European Economic Community led to the dropping of border tariffs between member states. Articles 100 and 100A of the Treaty of Rome[2] concerned themselves with the "approximation of laws" to even out competition between member states. This was necessary to facilitate the free movement of goods and people within the European Union. Article 118A was also added, which led to a Framework Directive[3] on the introduction of measures to encourage improvements in the safety and health of workers at work. Since the Directive's content was largely covered by HSWA74, the UK made the new legislative requirements "daughter" Regulations to the HSWA74. It introduced another level of detail into UK law welcomed by some as a helpful steer but claimed by others to obfuscate the Robens ideology that organisations should interpret what constitutes a suitable and sufficient response to risk.

Following the introduction of the European-led law, the HSE published Approved Codes of Practice (ACOPs) for these Regulations in an attempt to explain how to interpret them. These were launched with very clear opening statements saying that the guidance was not law, but, nonetheless, enforcers have used ACOPs as templates and courts use them as a mark of minimum requirements. Arguably, ACOPs introduced a form of prescriptive ruling, and this is one reason why in recent years many have been withdrawn. It is not well understood that it is still possible to implement a different but equally good risk control strategy to the one recommended in an ACOP. In the UK, HSWA74 still places the onus on the organisation to consider and implement their own rules to govern safety and occupational health.

The European-led Regulations did provide a clear benefit to UK legal processes. They brought in the language of risk management and provided the tools and techniques for identifying what the organisation should be actively controlling. It made explicit what was implicit in HSWA74. There is no mystery shrouded in this language. It outlines the requirement for good management (see Table 1.2).

The European-led Regulations do utilise imprecise language such as "fit for purpose", "suitable", "sufficient" and "adequate". This is understandable

TABLE 1.2

Tools and techniques brought by European-led regulation

Risk Assess	Identify, analyse and record significant safety problems.
Risk Reduce	Implement risk proportionate solutions to improve control and reduce the likelihood of accidents occurring.
Risk Monitor	Be aware that the nature and extent of the problem can change and that controls can work one day and not so effectively the next.
Consult	Involve all those with knowledge when making risk control decisions, particularly the employee exposed to the problem.
Inform, Train and Instruct	Tell people what will be done and give them the tools to do the job safely.
Health Surveillance	Keep an eye on the long-term future. Does the potential for harm also include a cumulative health effect?

since the rules have to apply in different cultures and countries, e.g. what is suitable and sufficient ventilation in Greece is very different to the northern UK. Many look to guidance to tell them what that means, believing the imprecise language to be unhelpful phrasing. We would contend that such interpretable wording is entirely in keeping with the intent of HSWA74 – that is, it is up to organisations to interpret the law and apply reasonable judgement as to how to comply with it in their individual circumstances. We explain how later in this book.

Bolstering confidence, certification and external recognition

As suggested, poor understanding and confidence in how to apply the law in many organisations generated a desire for external benchmarks. During the 1990s, there was a growing demand for certificated standards giving rise to standards such as OSAS18001 and BS14001 or for audit and accreditation from a nationally recognised body. More recently, this has continued with the publication of the International Organization for Standardization 45001 (ISO 45001).[4]

Whilst most standards start with the caveat that they should be adapted to the needs of the organisation, interpretation by those who do not understand this, or perhaps by consultants who do not fully understand how the organisation operates in practice, can result in the organisation overzealously applying them. The pursuit of certificated standards can very easily become a paper-chase exercise. Documentation provides an opportunity for auditing, perhaps by a recognised external body that may then heap praise for such a well-organised documented safety management system. The pitfall is that it can become a self-fulfilling prophecy: "I must be safe because my certification tells me which documentation I need, and my external auditor looked at my paperwork and said I looked safe". The reality in the work-place or between inspections can be very different. Organisations in this circular trap wonder why their incident rates don't really change despite the considerable investment involved in pursuing certification. You should drive what is needed in the safety management system. The system should not drive you. The external auditor may be a qualified safety practitioner, but it is unlikely that they will have a sound local knowledge that enables competent judgement of whether what you have is of any use.

A risk assessment is just a piece of paper, a record of understanding and thinking. A piece of paper does not control risk; it's the actions of people implementing physical controls and safe systems of work who reduce risk. We argue that burying the organisation in a bureaucratic nightmare of docu-mentation and due diligence checks might make you look compliant, but this does not necessarily reflect the reality in the workplace. Proving adher-ence to certificated standards and pursuit of recognised accreditation does not reduce risk in its own right. Using certificated standards and recognised accreditation only works if it leads to the full engagement of managers and employees who then collectively take action.

Who has responsibility? Where do occupational health & safety practitioners sit in all this?

In the UK, failing to comply with the requirements of the HSWA74 is a crim-inal act. Although prosecutions are normally brought against the organisa-tion, it will be the operational managers who will be standing in the dock giving an account of what they did or didn't do. The same can be said for civil cases too. Courts recognise that responsibility lies with the decision-makers. It doesn't matter whether occupational health & safety practitioners are called advisors, managers or coordinators; they rarely have the author-ity to overrule decisions made by operational managers. This central truth can lead to frustration when operational managers fail to listen to the prac-titioners. It's then very tempting for a frustrated safety practitioner to mis-quote law or standards as absolutes in order to force uninterested managers to take notice.

The legal duty to supervise is enshrined in HSWA74. Although it might be the occupational health & safety practitioner who designs a safety manage-ment system, it is the managers who are responsible for implementing it and ensuring that employees adhere to it. In the UK and in many other countries, wherever the managers may be in the chain of command, they are vicari-ously liable for the risks that subordinates manage on their behalf. Failure to demonstrate management and control of these risks would be viewed as negligent. Such responsibilities must be clearly stated in the organisation's Health & Safety Policy. A health & safety practitioner, whether a consultant or not, merely provides technical and legal advice so that managers can make more informed operational decisions.

A culture of risk aversion?

So, if HSWA74 is built on the philosophy that we should identify our own risks, interpret the law and apply it in a way that is sensible, proportionate and fits our specific local circumstances, why are so many organisations in the mindset that they never seem to get on top of their safety concerns? There always seems to be more that can be done. Pressure for continuous improvement comes from employees, union representatives, industry best practice guides, HSE guidance, consultants and litigation case precedent, insurance companies or by the media.

HSWA74 was written at a time when heavy industry, manufacturing and construction were dominant in the economy and where greater numbers of people lost their lives. In this century, automated manufacturing, leisure and the financial and service sectors have become the dominant forces in our economy. HSWA74 has been applied to wider and wider circumstances. The law of unintended consequences has come into play and has encompassed incidents with minor injury outcomes in low hazard workplaces too. It has helped create a culture of risk aversion where some organisations seem to have lost the ability to sensibly interpret the requirements of HSWA74 or feel less confident in doing so. This provides rich pickings for the sensationalist tabloid press.

One possible reason for the loss of confidence in the UK has been the changes to the civil procedure rules in 1998, the Woolf Report[5] in 1999 and the subsequent Access to Justice Act in 1999. Improving access to justice also created a free market for "no win no fee" lawyers that has unintentionally generated additional levels of fear in organisations. The number of litigation cases heard in UK courts did not significantly increase, but frequent advertising gave the impression that they did (of course it's difficult to ascertain the true number of cases settled out of court). The establishment of "no win no fee" lawyers laudably opened up the legal system to those who could not afford to participate before, but unfortunately it has certainly led to pressure selling techniques which has compounded the fear of being sued, particularly in small and medium enterprises. Lord Young's Report[6] in 2010 pointed out that following a successful legal suit, often it is only the insurers that stand between many organisations and bankruptcy.

It is worth suggesting at this point that insurance companies are quick to point out to a company what they believe to be intolerable levels of risk. Thus, they safeguard their exposure. We question whether they are quite so forthcoming with advice when an organisation has clearly gone too far and needlessly "gold-plated" their responses to the risk they face.

The continuous improvement of safety standards in the hope of reducing insurance premiums or in reaction to fear of enforcement or litigation is problematic. If this aim is pursued literally, it will inevitably take organisations to beyond what is reasonably practicable, towards the unreasonable "gold-plated" response to risk. There must be a point in time when problems are fixed well enough and the whole focus of the organisation should change from designing and implementing new controls to maintaining them.

At this point, it is worthwhile giving two examples to illustrate the effect of incessantly adding to a safety management system to demonstrate continuous improvement. A national retailer had over 750 risk assessments and associated controls covering its shop operations. It was more than could be remembered by managers. Once the ideas contained in this book were applied effectively, the amount of money being spent to resolve civil claims was reduced by more than a third in less than a year. By having a system that demonstrates what is important and, importantly, what is not, it enables management to concentrate their attention on what is really the most serious. In the second example, one of the largest social housing organisations in the UK had over 25 different policies covering health & safety issues. Each had its own set of responsibilities and procedures. It was so complex that busy managers could not work out just what it was they were supposed to do. In any case, if they did, they would not have time to deliver their operational role. Consequently, they ignored them and did what they thought best in the hope that it was.

Thankfully, the guidance to ISO 45001 addresses the point of continuous improvement more realistically. It reminds us that simplification is an appropriate response in the search to improve safety management and achieve enhanced performance.

One of the most interesting postings on the HSE website is the "ALARP at a Glance"[7] (As Low As Reasonably Practicable) page. In this section of the site, they discuss the application of reasonable practicability and the four most common fallacies of doing so. They provide advice for enforcement officers in a bid to prevent overzealous application. These are summarised in Table 1.3.

What this tells us is that we can and should stop attempting to control risk any further when we have done what is sensible and proportionate. It is not about giving in to the pressure from others but competently and confidently arguing what is reasonably practicable in our circumstances. That control at this point is safe enough. This will be discussed further in Chapter 2.

TABLE 1.3

Fallacies associated with ALARP

Fallacy	Explanation
Fallacy 1 – ensuring that risks are reduced ALARP means that we have to raise standards continuously.	• Improvements must be encouraged responsibly. • Just because it is possible does not necessarily make the judgement reasonably practicable. • If cost is grossly disproportionate to the benefit, then it is unreasonable. • Controls can be relaxed if the risk diminishes.
Fallacy 2 – if a few employers have adopted a high standard of risk control, that standard is ALARP.	• There can be many reasons why one organisation goes further than others, e.g. staff agreement, establishment of brand. • It is not necessary to force all similar organisations to adopt the same level of ALARP but to take a view about local circumstances.
Fallacy 3 – ensuring that risks are reduced ALARP means that we can insist on all possible risk controls.	• More than one way to reduce risk ALARP. • "Belt and braces" approach is not required. • It is not a requirement to reduce risk as low as possible. • If cost is grossly disproportionate to the benefit, then it is unreasonable.
Fallacy 4 – ensuring that risks are reduced ALARP means that there will be no accidents or ill health.	• ALARP does not mean zero risk. • Even with controls, risk will be realised sometimes. • You can only entirely eliminate risk by stopping the activity or removing the hazard. • An element of residual risk may have to be tolerated.

We would like to add two further fallacies to this list:

- Experiencing an incident does not necessarily mean that you have to change the risk control strategy.
- Having an accident that has not been covered by a predictive risk assessment does not necessarily mean that you were non-compliant.

Even with the most robust controls in place, it is possible that the circumstances on the day or a momentary lapse in controls can still result in an accident. It's uncomfortable but, nonetheless, a fact of life. It may well be right for an incident investigation to conclude that no changes to the risk control strategy are needed. Mistakenly failing to examine things in context can lead to unnecessary control changes, e.g. if a supermarket was able to show that the slip incident was one of six in the last 12 months, but during that period over a million customers had walked safely through the shop, then surely that is a vindication that the controls are working more than reasonably well.

By its very nature, a risk assessment is in part a prediction of what might happen in the future. Being a prediction, there is no certainty. Things could happen differently than predicted, or the accident could have been truly bizarre. It is possible that accidents may occur that have not been covered by a predictive risk assessment, but providing that the organisation does have a system for predicting risk and have monitoring systems to plug any gaps shown by experience, then it will still be compliant. We discuss this further in Chapter 3.

Compliance-driven versus risk-led approaches

Many organisations are compliance-driven in their approach, whether they acknowledge it or not. They believe that the more compliant they are, the less likely they will be to suffer enforcement and the more able they will be to fight spurious litigation. As discussed earlier, the trouble is that HSWA74 is not prescriptive. Its "daughter" Regulations contain imprecise requirements too. So how do organisations know when they are fully compliant? Figure 1.1 shows the different associated areas with undetermined boundaries reflecting that there may be more controls that can be added, more refresher training that can be done, better equipment that can be bought, etc.

When the European-led law came into force requiring organisations to complete risk assessments, many saw it as one more thing to comply with. Compliance was still central in their thinking, so risk assessment was just added to the list of things to do to be compliant. Some failed to understand that risk identification, assessment and control was actually the right thing to do to achieve precision and avoid needless bureaucracy.

What we argue in this book is that risk assessment should be placed at the heart of any occupational health & safety management strategy. If a

FIGURE 1.1
Compliance-driven safety.

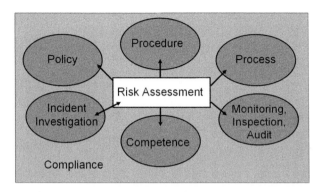

FIGURE 1.2
Risk-led safety.

systematic method for identifying risk is employed to ensure that things are not missed, there will be a certainty that the most significant risks have been identified. With that confidence, the findings of the risk assessment can be used to design precise policy, develop specific procedures, define competency standards and dictate precisely what is required in the training syllabus, etc. As Figure 1.2 shows, the world becomes simpler, finite and achievable when it is risk-led. Instead of health & safety being energy consuming as in the compliance-driven system of Figure 1.1, a risk-led system is actually energy efficient. A risk-led system identifies the limited times we need to act and provides the evidence for how much needs to be done. By identifying and proportionately acting to control the significant risk in its operation, an organisation will be able to demonstrate its legal compliance. Being risk-led is much more than simply having some risk assessments.

Future challenges

We are in a rapidly changing world. In developed countries, the workforce has an ageing demographic. People are now expected to work into their old age. Countries are increasingly less able to support retirement pensions and medical provision for those who have to retire early for medical reasons. Government and societal pressure is pushing organisations across the world to develop ways to keep people healthier and fit in the workplace for as long as possible. The pace of technological change is now so rapid that new technologies are out of date almost as soon as they go to market. These technological changes are driving changes to how and where people work. While they eliminate some health & safety risk, they introduce new ones. The nature of increasing competition drives constant organisational change and adaptation.

In the face of this ever-changing present and future, we suggest that developing and implementing a risk-led management approach is the most efficient and effective way of building a responsive and effective health & safety management system.

Summary

- The true meaning of health & safety law is often imprecise, but this allows leeway for organisations to argue what is reasonable for them.
- Imprecise understanding of health & safety law has led many organisations to look outside for direction or definition instead of inside for determination or inspiration as is intended.
- Organisations can develop an unhealthy reliance on what others say they need, even though those outside the organisation are unlikely to fully understand its culture, the way it operates and why and the issues it faces.
- A good risk assessment and risk management system will provide clarity for what controls and management systems are required, boost confidence and reduce needless bureaucracy.
- Simplification is an appropriate response to the desire to continuously improve.

References

1. Lord Robens (1972), *Safety and Health at Work: Report of the Committee 1970–72*, HMSO Cmnd 5034.
2. Treaty of Rome Articles. Accessible at: https://ec.europa.eu/romania/sites/romania/files/tratatul_de_la_roma.pdf.
3. Directive 89/391 – Occupational Safety and Health "Framework Directive", 1989. Accessible at: https://osha.europa.eu/en/legislation/directives/the-osh-framework-directive/the-osh-framework-directive-introduction.
4. International Organisation for Standardization 45001. Accessible at: https://www.iso.org/iso-45001-occupational-health-and-safety.html.
5. The Right Honourable the Lord Woolf, Master of the Rolls (1996), Access to Justice Final Report, Crown Copyright. Accessible at: http://webarchive.nationalarchives.gov.uk/+/http://www.dca.gov.uk/civil/final/index.htm.
6. Lord Young of Graffham (2010), *Common Sense Common Safety*, Crown Copyright. Accessible at: https://www.gov.uk/government/publications/common-sense-common-safety-a-report-by-lord-young-of-graffham.
7. http://www.hse.gov.uk/risk/theory/alarpglance.htm.

2

A toolbox of principles

In this chapter, we unpack and explain some of the legal concepts that must be understood if a risk-led safety management system is to be developed. These concepts are found in UK health & safety legislation, but they also form part of the legal framework in other countries. Where this is not the case, the authors have found that these legal principles are nonetheless useful tools that can be applied.

The chapter includes a description of reasonable judgement, competency, reasonable foreseeability, reasonable practicability and significance. The authors conclude that applying these principles in the context of the size and nature of the organisation's operation can justify health & safety risk decisions. They provide a foundation of argument that builds greater confidence in the organisation's risk management decision-making.

In Chapter 1, we discussed that the spirit and intent of the UK's HSWA74 is to require employers (managers) to predict and record significant risks and put preventative measures in place as far as is reasonably practicable. We pointed out that, without competent and confident advice, many organisations believe that others external to their operation must know better than them what must be controlled and to what degree. Organisations, therefore, have a tendency to look outside for guidance and definition rather than inside for inspiration and determination. So how can you grow the confidence to write your own rules? Quite simply, by improving your knowledge of the spirit and intent of the law and its legal mechanisms and how they can be used.

What do we mean by compliant?

During our time, as health & safety professionals, we have been frequently asked by the organisations we have worked for whether they are compliant with safety. No doubt this is an attempt to receive assurance, but is it the right question?

In Chapter 1, in the discussion of how safety law evolved in the UK, we described that initially law was highly prescriptive. Doorways had to be of a specific width, internal factory walls had to be cleaned and painted within particular intervals and even numbers of toilets and washbasins were dictated. With rules for just about everything, compliance was black and white. Following the adoption of the EU Directives, the move to more risk-based legislation meant that the notion of prescriptive rules had largely gone. Arguably, the notion of compliance should have changed too.

Compliance is now more subjective, and we argue that it is a good thing because it allows more ways to achieve the end goal. As we have described, however, it can leave people feeling uncomfortable. The reality is that organisations shouldn't be aiming to be compliant with safety regulation, but rather with the requirements placed upon us by that legislation. There is no requirement to have written risk assessments, but there is a requirement to consider what your business does and whether any of that is worthy of recording in a documented risk assessment. Theoretically, it is possible to be "compliant" and not hold a single risk assessment (although unlikely). Contrary to the belief of some, having a risk assessment does not demonstrate compliance unless and until the organisation can demonstrate it accurately, reflects reality on the ground and it is a tool used to drive the right behaviours.

Compliance isn't a binary response to meeting a prescriptive standard anymore. Neither is it just about having some risk assessments, controls and management systems. There are now two fundamental questions. Firstly, can we demonstrate that we have a (risk-led) system that recognises our obligations towards safety? Secondly, can we prove that it is delivering the necessary results?

Let's be reasonable

The application of reasonable judgement is fundamental to the success of any risk-led safety management system. It is a strong thread in UK health & safety legislation and so demands an explanation before we can describe the risk-led approach in later chapters. Managers are required to use reasonable judgement to foresee risk and then act to control it as far as is reasonably practicable. Being able to argue what is reasonable and what is not goes to the very heart of our civil and criminal law. Based on our experience, even when teaching this concept in different countries around the world where reasonableness is not included in local law, our students found the idea an incredibly useful concept they wished to adopt to some degree.

Whilst the test of reasonableness is now a cornerstone of statute law, its origin is to be found in common law in the UK and is related to the tort of negligence and the duty of care one individual has to others who may be affected by his or her acts or omissions. This is a complex and well-exercised area of UK law, but the following are some of the milestones in the development of this concept:

- *Vaughan v. Menlove (1837)*[1]: a reasonable person was defined as being an individual who "proceeds with such reasonable caution as a prudent man would have done under such circumstances".
- *Blyth v. Birmingham Waterworks Co. (1856)*[2]: reasonableness was discussed in terms of individuals meeting expected standards and if they failed to do so, they would be found negligent.
- *Donahue v. Stevenson (1932)*[3]: introduced the "neighbour principle" which extended the duty of care so that what was reasonable could now be examined in non-contractual arrangements, too.
- *Fardon v. Harcourt-Rivington (1932)*[4]: clarified that people must guard against reasonable possibilities, but they are not bound to guard against fantastic possibilities. If it is too unforeseeable to anticipate (i.e. bizarre or truly unpredictable), it was held that the Defendant has not breached any duty of care.
- *Hall v. Brooklands Auto Racing Club (1933)*[5]: the most quoted description of the reasonable person, i.e. "reasonable is, that which the man on the Clapham Omnibus regards as being reasonable"; in other words, the judgement of an average person going about their

business. In modern terms, what would the Defendant's peers say was reasonable.

- *Wells v. Cooper (1958)*[6]: gave further clarification that the test of reasonableness was an objective standard and clarified that it does not take into account the characteristic weaknesses of the defendant, i.e. it was found that a householder doing do it yourself (DIY) work must not fall below the standard to be expected of a reasonably competent carpenter doing the same job.

Being competent supports reasonable judgement

The central theme running through all of these cases and the hundreds like them is competence. This is demonstrated through an examination of the following:

- *Experience*: evidence is gathered by looking at the life history and background of the individual or decision-maker, e.g. where have they worked, what have they done?
- *Knowledge*: it is developed through training and can be evidenced easily by asking technical questions that the person should be able to answer. Knowledge is gained through experience too, not just by training or formal qualification.
- *Ability*: a little more obtuse but is an examination of skill, usually reflective of those things examined in appraisal systems, e.g. can this manager communicate both verbally and in the written word, or can this sheet metal worker resolve a problem found in the design, or simply can that person ride the bicycle?
- *Limitation*: probably one of the most important aspects of competency – knowing when things have gone beyond your competency and more specialist help is needed.

Competency is not a binary concept, i.e. either you are competent or not. Everyone develops his or her own competency on a daily basis by improving knowledge, gaining experience, etc. Competency is fluid and ever changing. That creates a legal problem; how do you know if someone is competent enough to make a reasonable judgement? When new employees are taken on, they rarely have zero competence. The advertising and interview stages try to ensure that people with the right skill, attributes and potential are taken on. New employees might not be able to fit straight into their job on day one; a period of adjustment and induction training is usually necessary. There comes a point that they are trusted to get on with it, i.e. they have reached the basic competency required for the role. From that day onwards, the employee will improve their competency by gaining experience for as long as they continue to do the job.

So when UK courts examine what is reasonable, what they are really asking is whether the person has the minimum competency expected to make such a judgement, a very good reason for organisations to think through and record the minimum competency requirement for each safety-critical role. Assessment and prioritisation of risk (see Chapters 5 and 6) is the foundation for deciding what must be in these records. They will include the knowledge needed, experience level preferred, the skill set needed and any formal qualifications if required. While sometimes this may reflect what is common in the industry, it should also reflect the standards that the organisation wishes to set for itself. It may be difficult to defend a reasonable judgement made by a company officer or worker without having this degree of definition recorded. This is not about limiting people in what they may attempt, rather it is about ensuring that if people wish to develop competency, they must do so (in safety terms) in a supervised environment. Learning by one's mistakes is not usually a basis for developing safety proficiency.

Reasonableness is also about local context

Where the risk is experienced and in what context is most important when deciding what is reasonable. Filing documents in an office and receiving a minor paper cut to the finger is not of any real concern for most offices, but if such an incident happened in a pathology laboratory where the technician is also handling diseased human tissue, concern would certainly be greater. Making a cup of tea at work is of no concern since it is a normal everyday task, but it is concerning for occupational therapists when teaching people with severe learning difficulties how to make tea for the first time. The context for reasonable judgement is who is doing what, with what, where, how, why and with whom? Context differs from one organisation to the next. Competitive organisations are structured differently, have different training programmes, different job descriptions, different business aspirations and make different decisions on the tolerability of risk. All these things and more will affect contextual judgements and decisions.

At a more basic level, the right people to decide what is reasonable and what is not are those involved with the job. Visiting enforcement officers may have a good knowledge of the law and associated guidance and may well have been in other organisations like yours, but what do they actually know about how your organisation differs from others like it, the local standards of competency required, the differing pressure experienced at different times or stages of the process, etc.? There has to be at least some gap in their knowledge. The only people with that knowledge will be the local managers and employees in the organisation. They are the best people to explain differences in management responsibility and arrangements peculiar to their organisation and method of operating. They are best placed to identify the nature and extent of the risk and decide what controls can be effectively applied.

It may be foolish to rely on external advisors to decide what is reasonable to foresee or what is reasonably practicable in your organisation. When we were consultants we used to train offshore gas platform workers in risk assessment. Geographical and logistic problems meant that the practical part of the course had to be done in a willing supermarket rather than on a rig. The result was a large number of risk assessments which predicted the potential to crash, bash and maim. This was a huge surprise to the supermarket chain whose quality risk assessments completed by competent employees painted a very different and more accurate low-risk picture; so why the difference in the two views? It came down to the fact that the gas platform workers were approaching the risk assessment of supermarket functions with their offshore mindset and not that of supermarket employee. The reason for their overzealous approach was the remoteness of their normal workplace where a simple leaking pipe could lead to catastrophic consequences when taken as part of the bigger picture. Every task on the oil rig was undertaken with absolute precision – right down to changing a tap washer. What they failed to take into account was the supermarket was a different setting and context.

Competency and understanding the context of the operation are the keys that unlock the ability to claim that the judgements made by the organisation are reasonable.

Using reasonable foreseeability is more accurate than a crystal ball

Thankfully, we do not have to think about and formally prepare for everything that is foreseeable. It would open up consideration of the trivial and the bizarre. We may foresee that aliens could attack us – there have been enough films from Hollywood on the subject – but is it reasonable to worry about such things becoming a reality? The vast majority of people would recognise that there isn't any evidence to suggest that this will take place. Based on present knowledge, it is an unreasonable and bizarre prediction.

As already stated, the benchmarking of what is a reasonable expectation is objective and is dependent on the level of competency needed. A court may only need a basic assurance of competency if it was related to an incident very much in the public knowledge and understanding; it may require the input of a true expert if the consideration is complex and highly specialist; or it may be anywhere in between the two. For ease of expla-nation, we suggest that this continuum roughly falls into three realms as shown in Figure 2.1.

The first realm is "what the public knows", i.e. general knowledge, not the more popular but less easily defined concept of "common sense". This is truly the realm of "man on the Clapham omnibus". We all should know and appreciate that which another normal member of society would know and appreciate, e.g. when it snows, smooth tiling in entrance foyers gets wet from pedestrian traffic and becomes slippery. Try saying "I never thought that could happen" to a judge!

The second or technical realm is "what others like us know or should know". This covers activities and situations that the general public at large would not be regarded as having a good enough degree of competency to make a reasonable judgement. They may know something of the subject but not enough, e.g. they may know that reversing lorries in a yard can have fatal results if a pedestrian is run over, but what do they know of vehicle route control systems in a distribution centre, or how banksmen operate in a construction yard? In such cases, the benchmarking would not be with the public but between transport managers or construction site supervisors. In this realm, our reasonable decisions should be made in the context of our own unique operation, but we must be mindful of what others like us are doing too so that we can demonstrate similar performance standards but not necessarily the same controls.

The third or expert realm is "what the experts know". This realm reaches beyond the levels of competency most would normally be required to have. Retailers may require some knowledge of lasers since they use them to read product barcodes, but they would not be expected to understand how the barcode reader is constructed and what laser light intensity would be regarded as unsafe. From their perspective, they ask the manufacturer for evidence that they have adhered to appropriate standards but not about the detail of how they complied. In this sense, the supermarket is in the sec-ond realm with technical expertise benchmarked with other supermarkets, while in relative terms the manufacturer is an expert.

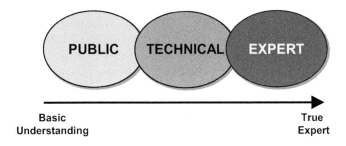

FIGURE 2.1
Realms of reasonableness.

The guiding light of significance

In the original Approved Codes of Practice (ACOP) for MHSWR99,[7] it is stated that insignificant risks could usually be ignored. This posed the question: what is the difference between a significant and an insignificant risk? This publication did not provide an explicit answer. It was vital that we got an answer to this; otherwise, we could have ended up wasting resources on identifying, assessing and controlling risk in a disproportionate way. We have since argued that reasonable judgement is the application of our competence in the context of the organisation's activity or business. It enables us to exclude bizarre and trivial risk. We can exclude bizarre thoughts like an irrational fear that a wild animal may break in and run amok through the office. We can exclude opening doors, walking up and down stairs or making a cup of coffee during a work break. That is unless we complicate things by requiring workers to repeatedly carry boxes up and down stairs, or if we take the normal to abnormal levels, e.g. a barista making 100 cups of coffee an hour at a café bar. As stated earlier in this chapter, reasonable judgement and arguments of significance must be within context. This concept is so fundamental to the ideas contained in this book that we discuss it further in Chapter 3.

In some organisations, arguably some physical hazards can be excluded from formal recorded risk assessments too like a loose carpet tile or a hole in the warehouse floor. If the organisation has a well-designed inspection and maintenance system that records the fault and prompts maintenance action, why is a risk assessment needed? The maintenance report and associated documentation is sufficient record to show the organisation both knows about and is dealing with the problem. Provided that local managers do something to control exposure to the hazard while they wait for it to be repaired, then the law is complied with.

We define significant risk to be all those risks that aren't insignificant. If we exclude the obviously trivial or everyday normal human activities and unreasonable (bizarre) predictions as insignificant and if we further exclude issues adequately covered by other management systems (e.g. maintenance removing physical hazard), then what we are left with must be significant enough for assessment. This approach certainly helps to tune the mind and greatly reduces the number of risk assessments an organisation needs.

One final thing to note on the subject at this point, it is our experience that many managers use reasonable foreseeability well when identifying how employees can be harmed when using a machine. It is an obvious prediction to say that a meat slicer used at a delicatessen counter has the potential to amputate fingers. The obvious control is to place a guard on the machine to prevent this from happening. That, however, is not the end of the story. What happens if you wish to maintain the machine by changing the blade, or if you need to clean the blade? In both cases, the guard has to be taken off. When using reasonable foresight, we must think beyond normal operational task and ensure that we encompass support tasks like maintenance and cleaning operations too. It is such tasks that often require the bypassing of normal safety protocols.

The pursuit of zero harm

Whatever legal framework is in place within different countries, they all seek to achieve the same thing, to put in place preventative measures to keep workers and others safe. Some have developed this intention into a pursuit for zero harm, particularly in countries or organisations that are still developing their health & safety legislation and changing cultural responses. This idea is of course an admirable one. To disagree with the concept of "let no harm be done" would be foolhardy indeed, however, it does not come without its considerations and caveats.

Initially it must be agreed upon what is meant by zero harm. Do we mean quite literally no injury or damage even at a cellular or psychological level; that would include a bump, knock or bruise, pinch of dust or a loud noise, or feeling you're having a bad day? Or, do we mean significant harm; the sort that might actually need some degree of medical intervention? Or, at the far end of the scale, are we purely considering fatal injuries? Eliminating fatal accidents would certainly be a good starting point for anyone.

Risk assessment must be included in the consideration of zero harm and potential contradictions that may result. We suggest that safety risk only exists where people are involved. A risk assessment has only two possible outcomes. As already argued, the first is that there is zero risk associated with the undertaking (insignificant) and the second is that there is some level of risk (significant). In order to make the process more flexible, we generally divide the area of "some risk" into categories of relative significance: low, medium and high. Most countries that utilise the concept of risk assessment have set some limits around this, typically only requiring the detailing or recording of significant risk and permitting insignificant risk to go unrecorded. We contend that most risk assessment methodologies use scales ranging from low to high and preclude zero. So, what is meant by zero risk? We'd argue that zero means exactly that. No "risk" at all. That is to suggest that there is no likelihood of any failure and no consequence either, even if there were to be a failure (the two are independent variables of course, as we discuss in Chapter 5). Achieving zero risk is possible, but only if we precisely define the undertaking in the assessment. Imagine making a parachute jump. There are a number of adverse things that could befall us between the jump and the landing. If we simply do not board the plane, then there is absolutely no chance of any of those things happening. If all flights are cancelled, then in reality there is zero risk. A level of risk cannot be attached to an undertaking that is eliminated.

An example of where zero risk (to people) can be achieved is where processes have been automated and humans excluded. In reality, maintenance and cleaning duties still have to be undertaken and some level of risk is thereby reintroduced. While the original task is now zero risk, other associated tasks may not.

That leaves us with the spectrum of low risk to high risk. As we argue more fully in Chapter 5, low risk does not mean no risk at all. Low risk indicates that in some way it's still possible to have an incident that may result in a degree of injury.

So, we contend that arguing for zero likelihood can only be done when tasks are eliminated.

Arguing for zero harm cannot occur when the organisation has risk assessments that suggest that a low risk exists, as there is recognition that some form of physical, physiological or psychological injury is possible and there is a discernable likelihood that such incidents could occur.

Reasonable practicability; knowing when to stop

Back to the consideration of reasonable judgement and now how it is applied to risk control decision-making. Lord Justice Asquith in 1949 helped to define what reasonable practicability is. In *Edwards v. National Coal Board*,[8] he said

> a computation must be made by the owner, in which the quantum of risk is placed on one scale and the sacrifice involved in the measures necessary for averting the risk (whether money, time or trouble) is placed on the other.

He continued by saying that "if it was shown that there was gross disproportion between them, in that the risk can be shown to be insignificant in relation to the sacrifice, then the Defendant need not act further to control the risk".

The application of competence to make a reasonable judgement is certainly used in this computation. It is not just about what is possible but using reasonable judgement to decide what is sensible and proportionate given the level of risk, the size of the organisation, the type of operation and the resources available.

For example, imagine a warehouse with both pedestrian and forklift truck traffic. There are a host of controls that are possible:

- Paint designated walkways.
- Put in a one-way system for the forklift trucks.
- Provide pedestrians with high visibility jackets.

- Put in zebra crossings where pedestrians cross the forklift truck route.
- Put in warning lights or traffic lights at crossings.
- Put signage up to warn pedestrians of the forklift trucks.
- Put signage up to warn forklift truck drivers of pedestrian crossing points.
- Put barriers up at the edge of walkways.
- Improve lighting.
- Develop a safe system of work for the driver.
- Develop a safe system for pedestrians.
- Train the drivers and pedestrians in the safe systems of work.
- Provide regular refresher training.
- Impose a key control system to prevent unauthorised use of the forklift truck.
- Regularly service the brakes on the forklift truck.
- Daily test the function of the brakes on the forklift truck.
- Train managers how to supervise the situation effectively.
- Remove people from the environment, i.e. fully automate.

All of these controls (and possibly more) will help to reduce the risk, but if it is a small warehouse with just the odd pedestrian walking through and the occasional use of a forklift truck, you wouldn't select everything off the list even though they are all possible; that would be unreasonable and disproportionate to the likelihood of incidents occurring. If it were a large warehouse with frequent pedestrian traffic and significant forklift truck movement, then you would probably implement the whole list; that would be a reasonable expectation.

Reasonable practicability is about balancing the level of risk and its operational significance against the cost of reducing it. The greater the significance, the more effort should be spent in controlling it. Fortunately, the law allows a line to be drawn when the cost of taking an action would be grossly disproportionate to the benefit that would be realised. Once again it relies on local competency to confidently provide the argument of why the controls selected by the organisation are sufficient in the context of the operation. Although the organisation should be mindful of what other similar organisations do to control similar risk, primarily it is the local knowledge and circumstances that should drive this decision. We can implement different controls and achieve a similar performance standard. The content of the risk assessment should describe what we do to control the risk and whether we need to do more. A quality risk assessment should provide the evidence and confidence for when to stop adding layers of control.

We will further explore the concept of reasonable practicability at various points in later chapters.

In summary, the key points made in this chapter are as follows:

- Compliance is not a binary state, it is a set of arguments based on reasonable and competent judgement underpinning risk control and safety management system decisions.

- Reasonable judgement is founded on the application of competency in the context of how the organisation operates. It is founded on local knowledge and understanding.

- Organisational and operational context must be taken into account when benchmarking control standards with external guidance. Being aware of what similar organisations do about the risk is advisable, but primarily control decisions are made based on local conditions and circumstances.

- Organisations should set and record minimum competency standards.

- We eliminate the insignificant by having a process that dismisses it and thereby defines what is significant.

- The aspiration for zero harm can be undermined by the content of our risk assessments.

References

1. *Vaughan v. Menlove* (1837), 132 ER 490 (CP).
2. *Blyth v. Birmingham Water Works Co.* (1856), Court of Exchequer. 11 Exch. 781, 156 Eng. Rep. 1047. Prosser, pp. 132–133.
3. *Donoghue (or McAlister) v. Stevenson* (1932), All ER Rep 1; [1932] AC 562; House of Lords.
4. *Fardon v. Harcourt-Rivington* (1932), All ER Rep 81.
5. *Hall v. Brooklands Auto Racing Club* (1933), 1 KB 205.
6. *Wells v. Cooper* (1958), 2 All ER 527.
7. Health and Safety Executive (2000), *Approved Code of Practice, Management of Health & Safety at Work*, London: HSE. Information accessible at: http://www.hse.gov.uk/pubns/books/l21.htm
8. *Edwards v. National Coal Board* (1949), 1 All ER 743.

3

Plan to finish

Many organisations struggle to answer the question: "Have you recorded all of the risk assessments you should have?" Without forethought, risks can be missed from the list or others needlessly included due to their insignificance. So how can organisations justify that they have captured all of their significant risks, with no exception?

In this chapter, it is explained how to plan for risk assessments and ensure that they are written in a justifiable order. It warns of the pitfall of using hazards to identify risks and describes other methods for identifying where significant risk may be found in the organisation's activities. The chapter finishes by describing how this risk identification preparation work may be used for maintaining the risk profile and help to capture organisational change within it.

During the second edition update for this chapter, it was discussed whether the subject of risk assessment had been adequately covered. Perhaps the concept was so commonly understood that this section might be deleted completely. In our view, such is the remaining confusion and misunderstanding of this subject that it was felt essential to leave it in. Risk assessment has been in common use for 30 years or more, but it still generates debate.

Before setting out on the journey

Unless you are a true devotee of the magical mystery tour, knowing the destination is crucial when planning your journey. That is just as true when embarking on a programme of risk assessment. There may be twists, turns and diversions on the way, but having sight of the destination is vital if steady progress is to be maintained. In 1992, the "daughter" Regulations to the UK's Health and Safety at Work etc. Act 1974 (HSWA74) were launched with the idea that if an organisation was already doing what it should to reduce the likelihood of incidents, it need to only document its position in order to comply. This was to include a demonstration that it had taken an inward look at itself to be sure that nothing had been overlooked. The challenge remains to capture the most significant and not to waste effort on risk assessing the trivial.

It took the Health & Safety Executive (HSE) and industrial bodies in the UK a few years before they published examples and further guidance on risk assessment. Unfortunately, this delay allowed confusion to take a hold and myths to generate which still pervade to this day. Although guidance was published about how to complete a risk assessment, there was little information on how to identify when one was needed.

The HSE's Five Steps to Risk Assessment[1] was an attempt to summarise the popular view of the time. Since its publication, there has been much criticism about this method, such as Spencer's article in the *Health & Safety at Work* magazine[2] and even from studies published by the HSE themselves.[3] Nonetheless, it is the method most commonly recommended by UK enforcement officers. We believe it has many flaws, including the requirement to list the additional controls needed when carrying out the assessment and once again when carrying out a review. It fails to be clear that sometimes no additional controls are necessary. Perhaps this is one of the sources of the myth that you never finish risk assessing. Can it be so wrong in a risk assessment to draw the conclusion that no further controls are necessary and that everything is fine just as it is? As we have outlined in Chapters 1 and 2, just because you can do something, it doesn't necessarily mean that you must. That just generates a needless "goldplated" response not required by either HSWA74 or its "daughter" Regulations.

The first part of the challenge is to identify what needs risk assessing first so that we don't end up further controlling things we don't need to or missing those things that we do. What we are advocating in this chapter is a way of defining which risk assessments are needed.

Don't reach for the blank risk assessment forms just yet

We argue that before conducting a risk assessment, it's essential that we determine whether one is needed. This is not helped by the fact that it is easy to confuse requirements in different areas of law, e.g. safety at work and public safety or health & safety and food safety & hygiene. A ban on homemade cakes at school summer fêtes was never part of Robens' vision, that's for sure.

Organisations need to be clear about where the boundaries lie in their operation. As a rule of thumb, you are only responsible for things you have authority and control over. It is not your duty to assess all risk because some of it will belong to others. Taking employees to a restaurant as a work reward shouldn't trigger a risk assessment. Not from you at least. You would be under the duty of care of the restaurant as a patron. It would be their responsibility to work out what they need to do to keep you safe (see also Case Study 3.1).

CASE STUDY 3.1

Hospital A requested that hospital B send their staff to run a specialist clinic once a week since they did not employ that specialism themselves. It was agreed that hospital A would provide the room and purchase the equipment, and hospital B would provide the expertise. So who was responsible for what?

- *Workplace safety*: hospital A since they owned the room and therefore were responsible for maintaining floors, walls, windows, etc. They also retained the responsibility for the safety of patient's journey to and from the unit on site.
- *Equipment calibration*: by default hospital A since they owned the equipment; however, by contractual agreement, this could be transferred to hospital B.
- *Clinical decisions*: the responsibility of hospital B since it was their specialist staff making these decisions independent of hospital A.

Risk assessment versus assessment of risk

In the UK at least, the term risk assessment has been retrospectively applied to other subjects in the health & safety field. In the 1980s, a new legal requirement to consider chemicals and their harmful effects was launched. At the time, there was a new requirement to produce CoSHH (Control of Substances Hazardous to Health) assessments. Somehow in more recent times, these became corrupted to be CoSHH risk assessments. Similarly, in 1992, there was a need to perform manual handling assessments. These too became corrupted and falsely described as manual handling risk assessments. So what's the problem? It may be semantics, but if the term risk assessment is given formal legal status and definition by both the criminal and civil justice systems, then there are implications for the presence or lack thereof. This may be considered a minor point, but it is yet another example of where "mission creep" has entered health & safety thinking.

Making a start

More heads are better than one, so it is recommended that a small team be assembled. The team's size will be related to the size and complexity of the organisation. The members will need to be well informed, mobile and with the time and enthusiasm for the project. We would also recommend that employees are invited as well as managers. These core members don't need to be experts in everything, they just need to have a good understanding of the organisation and know the "right people" to call on for information when needed.

The role of this team is to identify "the tasks that we do". They should be able to produce a comprehensive list of all tasks completed in the operation. Their discussion must include the more obscure activities that may only be performed infrequently, e.g. some foundries only have a major shutdown once every three years when some really "scary stuff" goes on. So, the team shouldn't just capture all of the everyday tasks but look for the reasonably foreseeable abnormal undertakings too. Popular experience shows that it's often an infrequent or a complex task that is prone to incident. We are not after bizarre things (see Chapter 2) but just all of the things that potentially we would have to be honest about in hindsight and agree were reasonably foreseeable.

Why focus on tasks – what's wrong with hazards?

The description of a task also provides the context for the risk assessment that follows. Consider a typical training room with desks, chairs, a television and a projector, perhaps even a coffee machine in the corner. When you list the things with the potential to cause harm, we may come up with trailing cables, electrical sockets, corners of desks, hot drinks, etc. How many of these hazards would you be worried about if it was a "room regularly used by a trainer to train an adult workforce?" – probably none of them in truth. What if we changed nothing in the room and started to use the training room as a playroom for very small children? We would no doubt argue that the number of hazards doesn't change, but how they would be potentially interacted with does. Think about these two task descriptions – "using the room for adult training" or "using the room as a crèche". The task gives the context for assessment, indicates whether controls are necessary and if so, what controls should be implemented.

This is supported by HSWA74 and associated guidance, which require us to understand and be in control of our undertakings. This has a direct correlation with "what we do" as an organisation. Largely, we pay people to do things that when joined together make up our organisation's operation. So, if we know what people are doing, then surely we can work out if they (we) are doing it well. This might sound obvious, but if people didn't do anything, it couldn't go wrong and people wouldn't be hurt as a consequence.

So, there must be something about performing tasks that is the link. Participating in a task that leads to an adverse interaction with a hazard is what leads to injury, i.e. someone having an accident. What we need to do is to prevent the incident. If we are to identify these potential events via risk assessment, then it makes complete sense that our starting point is the tasks that people do. Convention dictates that we regard hazards as the things with the potential to cause harm. It would be a great deal of work to eliminate all hazards in the workplace, which would be impossible and even undesirable.

A good way to think of risk assessment is as follows:

- What could go wrong and why?
- What do we do to prevent it going wrong and do we do those things well enough?
- What will we do if it still goes wrong and do we have that capability?

CASE STUDY 3.2

A leading supermarket used hazard spotting to identify its safety risk. Reported accidents and also inspections by local enforcement officers added more and more potential hazards to the list. It got to the point where supermarkets with a petrol filling station had in excess of 750 hazard-based risk assessments. This not only created a needless bureaucratic nightmare but also the system lost all credibility with managers. It became a "tick box exercise" and bore little reality to what managers and employees were doing on the ground. Managers could not and did not manage risk because it wasn't clear what they should focus on.

The supermarket changed to a task-led identification system for risk. Although 110 risk assessments were still required for this type of store, much of these were department specific. In reality, section managers had around 6–12 significant task-based risks to manage. This change enabled more focused and accountable management. The result was a quick, significant and sustained reduction in both accident numbers and litigation claims.

Walking around the organisation with a clipboard looking for hazards does not provide robust results. Some hazards may be present but you don't need to implement further control of them. It is true that someone may fall off the roof when working up there, but if it is contractors' and not your employees that work up there, then it's their risk to control and not yours. You would just have to satisfy yourself that they had thought it through and were planning to implement appropriate controls, through dialogue and contractual arrangements in order to manage your liability. Of course, not all hazards are visible. How would you spot electricity, wind and radiation? Not all hazards are present all of the time and can be transitory, e.g. a reversing lorry once a day. Hazards on the ground such as blossom, leaves, snow and ice are of course seasonal.

Being truly systematic

Compiling a list of the more physical aspects of what people do for us takes a little practice and probably brainstorming is best avoided as by its very definition, it isn't systematic nor thorough. What is needed are some anchor points that are finite and easily identifiable. A good place to start is to consider people, equipment and locations (see Figure 3.1).

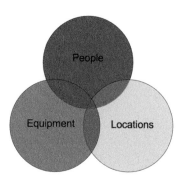

FIGURE 3.1
Identifying tasks.

We will need to consider all of the circles in Figure 3.1, but often there is one that makes a sensible lead. It depends upon what type of organisation you are, e.g. an outdoor adventure centre would favour location and a manufacturing company would favour equipment.

You will need to find the most logical way of splitting up the organisation; there is no one right way. A typical supermarket is a simple example and we might start by using **locations** as a way of dividing the world into manageable pieces:

1. Car park
2. Shop floor
3. Warehouse
4. Office areas
5. Rear yard

There may be merit in breaking these areas down into smaller areas:
Rear yard:

- Unloading bay
- Vehicle parking bays
- Refuse collection area
- Empty roll cage storage area
- Smoking shelter

Within those bite-sized areas, we have now the opportunity to look at the equipment that exists there. In many larger organisations, this list may turn out to be fairly common but don't be lulled into thinking that doing this at one site will mean that you have captured everything for everyone. We must be very wary of promoting generics; in the aftermath of an accident, many multiple site organisations have been criticised for generic approaches that did not sufficiently accommodate the local conditions at the site. What we must achieve is a task list that every location can use to pick off the tasks from the list that apply in their location and challenge when they believe tasks are missing. The team must watch for managers who declare "but we don't do that here", their surprise is palpable when they realise that in fact they do.

So the equipment list might look like as follows:

- Refuse collection area:
  - Baler
  - Compactor
  - Pallets
  - Waste oil tank
  - Roll-top bins

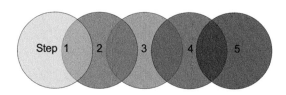

FIGURE 3.2
Identifying process tasks.

Now, we are starting to make some progress. With the consideration of "people" laid over the other two lists, we should be seeing a picture emerging, e.g. a warehouse operative – refuse collection area – baler. This forms the stem of a task title.

Another very useful way of getting the list together is to consider if there is a process involved. This approach isn't really an alternative to the method described above, but could be used as well. Where operations are perhaps more linear, e.g. in a police custody suite where prisoners arrive, are booked in, details taken, searched and then taken to a holding cell, the circles represent each of those quite well-defined operations (see Figure 3.2). Someone, however, will still make for a prisoner a drink, take them their meals and clean out their cells, etc., tasks which are not part of the core process but are supportive of it. Almost all organisations will have a need to use a mixture of the two approaches. Whether the process in Figure 3.1 or Figure 3.2 is favoured, once the raw task list is written, it is as well to go through it one more time and consider: is that a core task and are there any supporting tasks we have missed?

Describing tasks

At this point, actual tasks will start to emerge: in essence, someone, somewhere using something. Tasks must contain a verb, e.g. warehouse operative driving a forklift truck in the yard, warehouse operative *stacking* empty pallets at the end of aisles, delivery worker *moving* furniture into people's homes. The hazard description can be incorporated too if it helps with clarity, e.g. delivery operative moving *heavy* furniture into people's homes, electricians working on *live* electrical systems in remote switch rooms. This gives the reader an insight into what they are about to read.

Careful thought is needed, something that at first may seem to be a sensible task for risk assessment may not be. Replenishment of shelves in a supermarket is not a task, it's a general undertaking made up from smaller components such as handling cages, unpacking stock, cutting open boxes, etc. Similarly, the failure to act appropriately or implement a control is not a task. Failure to follow the speed limit by driving a forklift truck too fast is not a task in itself; it's a task being done badly which increases the likelihood of an accident occurring.

When looking at **equipment**, it is necessary to consider all of the ranges of people identified in the other circle to determine who does what with it. A warehouse operative will be doing a very different thing with the baler than the maintenance technician. Operating the baler is only one of the associated tasks with it; we must consider rethreading baler strapping, oiling moving parts and clearing a blockage in the baler chamber. As noted in Chapter 2, cleaning and maintenance tasks often require an employee to bypass normal safety controls to be able to do it, e.g. you have to take the guard off a meat slicer before the blade can be changed or cleaned.

When looking at **people**, we must consider all of the people who will come into the picture, but once again there are limits. Customers do not operate supermarket warehouse equipment, nor should they be anywhere in the vicinity when it is being used.

The **location** category is an obvious one to think through, but it's easy to miss the obvious such as main entrances, reception areas, terraces, service tunnels, etc. Again, being systematic is the key. Care is needed to ensure that rooms are not risk assessed, that requirement is met by safety tours and inspections rather than risk assessments. What we are interested in is what tasks take place there? Whilst making a start sitting at a desk is a good idea, it is not a desktop exercise and there is no substitute for getting out and watching what employees are doing.

As discussed in Chapter 2, other processes better deal with some things rather than risk assessment. Management systems such as safety tours or inspections are more simple tools. Not only should these walkabouts be looking for failed physical controls (e.g. missing guard) and poor behaviour (e.g. running through the shop floor) but also for physical hazards (e.g. a hole appearing in the shop floor forming a trip hazard). Such things are quickly managed and the solutions are usually obvious. Adopting a risk-led management system must be supported by the existence of maintenance and supervisory management systems too.

Reducing the size of the list

It is true that if we wrote a list of every single task that required the input of an employee, we could write an enormous list. While in theory anything could be allowed onto the list, in reality there have to be some editorial rules. Sharpening a pencil or opening a letter in the office should really not be allowed onto the listing even though we are being open-minded.

Task numbers are reduced by grouping very similar actions together, e.g. in maintenance, we could have assessments for the use of hammers, saws, screwdrivers, pliers and craft knives. This can be summarised simply as technicians using hand tools. The reasonably foreseeable injury arising from the use of any of these for their intended purpose would surely be very similar – cuts and bruising. No one is likely to inadvertently saw his or her own limb off with a handsaw.

More on significance

Once the task listing is compiled, it is necessary to decide which of the tasks, if any, actually require risk assessments. The team must discuss the significance of the task to the organisation. A word of caution here, do not attempt to guess the level of risk at this point; that is the function of the risk assessment part of the process that follows. It is very easy to confuse significance with some notion of the level of risk. Separating the significance of the task and the significance of the risk is a discipline that needs practice. The following questions will help to assess significance for the tasks in your organisation:

- How long would we continue to function if we couldn't do that anymore?
- Is this one of our core functions or something that we do that enables a core function to run efficiently and effectively?
- What would the public, or indeed a Jury, think if we didn't have an assessment for that particular activity? (Be careful not to let "the fear factor" to creep into your discussion.)
- What defence could we put forward for not having it?

For example, a department store may require staff to occasionally make a display bed in a shop window. Even though a muscular injury could be suffered, we would argue that this task was no more and no less than any person would do in their everyday life, therefore the task would not require risk assessment. It would be a different conclusion if we now put making beds into the context of major hotel chain. We would rightly argue that for hotels this was not a significant part of housekeeping duties.

The arguments of significance will enable you to put the task in one of the four boxes shown in Figure 3.3. Reasonable judgement will indicate that the box on the top right will be tasks that must have a risk assessment, e.g. reversing delivery vehicles onto the dock across a public footpath. It

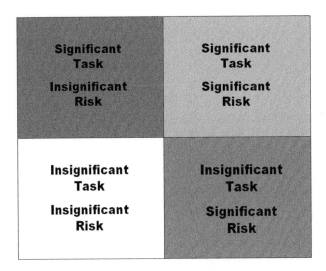

FIGURE 3.3
Task and risk significance categories.

is the category in the model where you would start your risk assessment programme.

If the task is significant and the risk not (top left box), a closer examination during the risk assessment phase that follows may well reveal some tasks which already have robust controls in place that reduce the likelihood of incidents. In such cases, you may wish to review contingency.

Insignificant task but significant risk (bottom right box) needs some careful focus. The task may be insignificant to your operation, but there is a potential for a serious incident, e.g. employees don't go on the roof, but someone has to clear the guttering. Typically, this might be an area where contracting out to organisations with better competence and equipment may be prudent. By suggesting that contractors or specialists should be brought in, we are not abrogating responsibility for risk ownership, we are merely suggesting that the team must consider who will bring in the correct skill set to deal with them. What is an abnormal or a dangerous activity for you will probably be normal activity for others.

Insignificant task and risk (bottom left box) will contain trivial issues where there would not be any significant findings if a risk assessment were produced. Essentially, they are activities that simply fall well below the radar in terms of safety, e.g. making hot drinks or using the microwave in the snack and restroom. Placing these on a separate list demonstrates that a mental risk assessment took place and that there were no significant findings and therefore no recorded assessment is required. This is far better, neater and easier than using a risk assessment methodology that produces a physical assessment document with the rating "insignificant" recorded on it. It suffices to just keep a list of the tasks you have placed in the category as proof you considered and then dismissed them.

Now we have a plan because clearly our priority is the red list of tasks. Remember that in terms of risk assessment, we are looking to record our significant findings – significant in this context meaning "worth writing down".

Once the lists have been roughly cut in this way, the team can sense check the lists by reading them through and discussing what hazards are associated with each task. This simple exercise – a rudimentary verbal assessment – is enough to ensure that the tasks are in the right category.

Ordering the list

The process described above will create a jumbled list of tasks. This can be confused and provide the impression that it is an illogical list, e.g.:

- General use of office equipment
- Refuelling of forklift trucks
- Gritting
- Unblocking the compactor
- Moving nests of cages
- Topping up batteries
- Unloading containers

To help managers to engage with the process, we suggest that those assembling the task list place the tasks into categories. We could use process to order them, e.g.:

- Reversing goods vehicles onto bays
- Operating dock levellers
- Opening roller shutter doors
- Unloading cages

We could list them by location (but they would still need to be ordered within those locations), e.g.:

- Yard
- Warehouse
- Battery changing room
- Offices

By core task, support task and miscellaneous tasks:
  Core, e.g.:

- Unloading vehicles
- Putting away stock
- Picking into cages

By support tasks, e.g.:

- Changing FLT batteries
- Repairing damaged cages
- Assembling post pallets
- Operating the baler

By miscellaneous tasks, e.g.:

- Replacing blown lamps
- Clearing snow
- Stacking pallets

At this stage, the risk assessments have not been completed, so the tasks cannot be listed in risk priority. In any case, this is what the risk register is for (see Chapter 6). It is possible, however, to use reasonable judgement and list the tasks in order of operational significance. Again, this logic helps to engage operational managers. It reflects that even though the risk level may change by imposing controls, the significance of the task to the operation may not.

Once the lists have been arranged in this way, identification code numbering can be imposed. An identification code provides a useful shorthand when referring to risk assessments in internal correspondence, incident reports and procedures. Doing so means that the full description of the risk need not be explained each time. When numbering the list, leave some gaps so that new risk assessments may be added in the future, or better still use a prefix to indicate which category it is in. For example, CT001 would refer to core tasks, unloading vehicles; and MT001would refer to miscellaneous tasks, replacing blown lamps.

Maintenance and future proofing

The reality is that even with the most systematic approach, there is the potential to miss a task. It does happen. There may be additional tasks to add to the list as time goes on to reflect changes in operation and equipment, but these should be few in number and easy to manage once part of a regular review process.

If you had mistakenly forgotten to include one significant task in your list, you will have a mitigation to argue. It would be possible to show the assessments that you do have, the plan to write the others not yet completed and the ones that you decided not to have. When challenged, at the very least you can say – "OK, but see how hard we looked?" If you can also show that you periodically reviewed the content of your task list, your mitigation is improved still further.

One last point: the task list becomes the index for the risk assessment file. It becomes a very useful tool when discussing things with an enforcement officer or auditor. When they ask to see your risk assessment for that hazard over there, the task list can be used to help explain whether or not anyone is doing a task which brings them into contact with it. If there is such a task, the task title will identify which specific task-based risk assessment needs to be explained.

Identifying tasks for assessment in this way is the first vital step in a risk-led management system that drives evidence-based actions. It ensures that you look in the right places where you may have safety issues. Understanding the concept of significance is vital to the success of the task-based risk assessment technique. The development of task titles ensures that the way you are looking at what you do, with what and where is kept in a meaningful context. Any manager or employee reading a task title should immediately be able to understand its relevance and start to understand why an assessment is needed. Without such a technique, you will inadvertently waste your effort by including the assessment of trivial or bizarre risks in your documentation.

In summary, the key points in this are as follows:

- Task analysis, reasonable judgement and understanding operational context and significance provide opportunities to identify what risk assessments are required.

- Tasks identify whether a risk assessment might be needed or not. Someone has to be doing something, with something, somewhere to come into contact with a hazard.

- Start with a well-briefed, motivated and empowered team when writing task listings. Data gathering requires mobility and knowledge of both the organisation and the workforce.

- Tasks must have verbs in them – they are the things that we pay people to do in our operation. It is these tasks that we want to prevent from going wrong which can ultimately lead to an incident. Hazards are too many and varied to be useful as a starting point and not all of them are relevant.

- Start writing assessments of significant tasks. Don't write assessments of the insignificant first, as the programme will wither from lack of interest.

References

1. Health and Safety Executive (2011), *Five Steps to Risk Assessment*, London: HSE. Accessible at: http://www.hse.gov.uk/pubns/indg163.pdf.
2. Spencer, D., (2012), *Unsteady Steps, Health & Safety at Work*, August 2012, page 18. London: Lexisnexis Butterworths Journals and Magazines.
3. Health & Safety Executive (2006), *An Evaluation of the Five Steps to Risk Assessment*, HSE Research Report 476, London: HSE. Accessible at: http://www.hse.gov.uk/research/rrpdf/rr476.pdf.

4

Risk assessment friend or foe

Frequently organisations are criticised for having risk assessments that are full of mistakes or possessing ones that lack clarity because important information is missing. In this chapter, the ingredients of risk assessment are teased apart to help illuminate the arguments and techniques that will help to assure the production of quality documents. It includes ideas on using the right language and the importance of establishing context. It contrasts generic and localised risk assessment. It discusses the meaning of suitable, sufficient and adequate and how they relate to the production of quality risk assessments. Finally, it introduces the concept of likelihood and when the content of a risk assessment is mature enough to only need maintenance rather than the needless adding of more controls at each and every risk review. These ideas are explored further in later chapters.

As previously suggested, the subject of risk assessment is a hotly debated topic. There are many views and many methodologies to choose from. Some are more suited to certain operations than others and some are very specifically designed. We will narrow the field in this chapter and talk only about general risk assessment. That will be more relevant to most readers. In our view, we contend that there is a difference between a risk assessment and the assessment of risk. The former is a more formal document as we shall go on to describe later.

Before talking about the elements that we suggest should be in a risk assessment in Chapter 5, it is necessary to discuss nine of the common pitfalls that befall the unwary. In this way, we hope that the reader may find information that will help them to improve the method they presently prefer.

Pitfall 1: Is the language right?

The word risk is often used to mean different things and can lead to confusion, i.e. "there is a risk that the risk of an accident could lead to a risk of an injury". We must accept that in general speech, there will always be variance both in use of the word and in the way the listener hears it. Risk is used to describe *danger,* to describe *variability* or to describe *uncertainty.* A weather reporter on the television may report that there is a risk of snow tomorrow. Whether he actually meant that snow could create *hazardous* conditions or whether he meant to reflect on the *likelihood, probability* or even *possibility* of snow is open to individual listeners to work out for themselves. The word "risk" is often used interchangeably with all of these words. What is plain therefore is that organisations must decide how they will define risk and become disciplined with the use and meaning they intend.

This matters a great deal. If your boss has a different perception of what risk means to you and uses different language to describe it, when you report a risk problem to them requesting additional resources, their differing concept of risk could colour their judgement leaving you without the help you need and both of you legally exposed (your boss is vicariously responsible for the risks you manage on their behalf). Equally, a senior manager acting on a risk presented to them, which in truth is not significant, would be wasting valuable organisational resources. The discipline of language is therefore vitally important when risk is reported between departments or management levels, particularly when describing hazard or the likelihood of incidents occurring.

Pitfall 2: Who is the risk assessment for?

As we described in the previous chapter, tasks provide the context for risk assessment. It's what people are doing that exposes them to hazard. It's how they implement controls while doing the task that keeps them safe.

We have argued that a risk assessment must begin with a suitable description of the task, but one crucial question must be answered: "Who wants what from the risk assessment?" It's our experience that employees are seldom interested in the content of a risk assessment. It's not a training document and presumably employees already know what they are doing. The problem here stems from the common misperceptions that we need them purely for legal compliance or that risk assessments are written for employees. True, employees have a right to have findings communicated to them, but employees benefit more from understanding the controls that spill out from its conclusion rather than the content of the risk assessment itself.

The reality is that risk assessments are of far more use to managers as an information source recording the analysis and as an aide memoir of controls. Managers do not always understand the detail of what their employees are doing or how they are doing it, especially if the manager is remote from the activities of the shop floor. It is impossible to effectively manage that which you don't understand; the task-based risk assessment must fill this knowledge gap. This matters if a control requires a high degree of supervision when relying on the operator alone would prove insufficient, e.g. ensuring employees wear high-visibility jackets in a vehicle yard.

Pitfall 3: Is the context clear?

Put simply, receiving a paper cut in an office is not the same as getting one in a research laboratory working with human pathogens. Understanding context is essential. It is partly set by the description of the task, e.g. moving pallets of waste cardboard across the vehicle yard using a manual pump truck. As we have argued in Chapter 3, the description of the task must set the scene so that the reader starts to think about who is doing what, with what, how, when and where. It is clear that we are not talking about moving inside the warehouse but outside in the yard. It is clear that it is a potential manual handling issue. A local manager will immediately be thinking of the route, the frequency, the use of the yard by vehicles, the condition of the ground in the yard, when waste cardboard is produced and how much of it there is, etc.

The risk assessment would be better still if there is more information that expands on these points and includes, where relevant, best practice or any industry standards that apply. Such a wealth of information would provide a visiting enforcement officer or auditor with much-needed understanding of why the task has to be. It may also be useful in explaining context to record less demonstrable elements such as custom, practice and culture. In our preferred risk assessment method, we call this section of the task-based risk assessment the synopsis.

The synopsis plays a major part in providing consistency throughout all versions of the risk assessment across the organisation, as we will discuss more fully in Chapter 5. There may be some differences of how a task is performed from site to site, but how differently could you describe using a stepladder, replacing light fittings or driving a forklift truck? This again underlines our point that a well-conceived and -worked-out task list is absolutely invaluable to the management of safety risk. It provides common agreement on the context of the tasks being completed.

Managers often change roles when covering leave, as part of acquiring experience or even following "realignment" of the workforce in challenging times. Why we do a task, or why do we do it in in a particular way, may not be clear to new or seconded managers. What does a manager need to know when they arrive in a new department? They will doubtless understand the basics, probably more than that if they have only moved internally, but they will be missing the history, background and local knowledge held by their predecessors. The synopsis on a risk assessment is a narrative that provides sufficient explanation to someone who isn't familiar with the task or the local control of incident likelihood. This also proves to be invaluable when explaining things to an enforcement officer, auditor or when defending the organisation in court by helping those unfamiliar with the operation to understand.

Pitfall 4: Generic versus locally specific risk assessment

It is a generally held view that generic risk assessments are not necessarily a good thing. Probably it is a result of a number of legal cases where the generic risk assessment has been shown to inadequately reflect specific local circumstances relevant to the incident, e.g. it did not take into account that unlike other locations, this particular yard slopes which is one of the reasons why the trailer ran away. The one good point about a generic risk assessment for large or complex organisations is that when sitting in head office, you know exactly what the assessments do and do not contain. So if not through generics, how do you ensure quality in assessments for a large or complex organisation where there could be many assessors, all locally creating different documents? After all these risk assessments are not only used by local management but must also potentially defend the business.

One solution to managing this challenge is to part-write and release a risk assessment as a template for local managers to complete. They must not be given permission to change what is written in the template, but they must be trained and allowed to add comment to make it locally relevant. Most importantly, managers must be able to challenge the template with those who created it. This allows for very effective central monitoring and is a key component for audit and quality assurance.

Pitfall 5: Suitable, sufficient and adequate

If we go back in time and look at the origin of risk assessment (see Chapter 1), it was clear that the philosophy was "make sure that you have a good look and come up with improvements if and when necessary". There are risk assessments that tend to try and identify each and every hazard. Such an approach is often driven by the fear of litigation. Risk assessment shouldn't be about justification and covering every conceivable hazard, which has been the driver for some safety professionals in the UK. They do so not because it is really helpful to do so, but because there is a misconception that they are duty bound to do so. We must filter out the trivial and bizarre issues and stick to those tasks and risks that are significant.

The law only requires significant findings to be recorded in a risk assessment, but what does that mean? Police officers deployed in crowd control may potentially face petrol bombs, missiles and all sorts of other weapons. They could also trip over a kerbstone or get sunburnt as well: two hazards that in this context are just not as significant as the rest. Although they may be mentioned in a risk assessment, they must not be the principal focus of it.

When predicting the reasonably foreseeable incident, the principal hazards will be quite apparent to the assessors. In many cases, there will be one principal hazard which when addressed will often nullify other hazards associated with the task. Noting multiple hazards is fine, but it is usually not necessary to analyse all of them further. Typically, the hazard with the potential to inflict the most harm in the most obvious way would be the one to carry forward, as long as the assessor hasn't strayed into the bizarre of course – e.g. when marching out to police the crowd, we could all be struck by lightning!

Predicting what could go wrong and why gives any manager reading the risk assessment a prompt for what they need to look out for, e.g. employees could be forced to overreach and fall whilst on a ladder due to obstructions in the aisles that prevent them from positioning the ladder correctly. This would be a very clear steer for a manager.

Pitfall 6: Predicting injury and rating hazard

Now that we have an understanding of what people are doing and what can befall them if it goes wrong, hazard needs to be quantified in a useful way – just how badly could they be hurt. Not all assessments have this step, but making a prediction of injury allows the application of some form of rating. There are those safety professionals who simply abhor any kind of scoring system. If you are presently in that camp, you'll see in Chapters 5 and 6 why scoring can be extremely useful.

Firstly, in predicting the injury, there is no need for medical training; nothing needs to be that complicated. A reasonable prediction of the injury extent given the incident predicted is all that is needed. Injuries could then be categorised on a scale. Some risk assessment methodologies propose a scale of 5 or more. We argue that if the scale includes any more than three categories, it produces difficulties with labelling and usage. On a five-category scale, common divisions are multiple deaths, death, major injury, minor injury and insignificant injury. That does seem like altogether too many choices and the last choice – insignificant – does beg the question that if we are to record only the significant findings in a risk assessment, having the choice of insignificant would seem to defeat the purpose.

Our preference is most certainly a three-category scale. The choices are then really simple: good, bad or somewhere in the middle. That is much easier to understand. It's either in one of the extreme categories and if not, it's in the middle. This is explained more fully in Chapter 5.

Pitfall 7: Lost time confusion in hazard prediction

There are assessment methods that use a notion of lost time or even the inclusion of regulations on reporting injuries in the definition of hazard scales. This is in our view unhelpful as it tends to confuse people and in terms of making a reasonable prediction can be influenced by local circumstances and culture. Is a twisted ankle from a fall likely to result in a reportable injury? It depends where you work. Some people might take two weeks off work and others might come into work and carry on. The injury outcome from a fall would be the same but the evaluation would not be.

Pitfall 8: Misunderstanding likelihood

Where people often seem to stumble is in the understanding that likelihood is the *likelihood of* something happening (incident). As we propose in Chapter 5, using a simple graph of hazard consequence versus likelihood, the laws of mathematics dictate that each axis should be an independent variable: therefore, likelihood cannot be the likelihood of the injury. In simple terms, likelihood is the possibility of controls failing in a way that creates the right conditions for the predicted incident to happen. Imagine employees walking across the yard, some without donning their high-visibility jacket. This time they make it without incident, but the controls have still failed. Even though they were not run over, it is still possible that drivers did not see them. This may seem a pedantic point, but do we have to wait until an employee is run over before we act, or do we challenge them to prevent it going wrong in the future? Likelihood, therefore, is about adherence to and maintenance of controls and their effectiveness over time. A likelihood factor of anything other than zero doesn't mean that it's inevitable that an incident will occur. It's an estimation of the realistic chances given what we know about the task and the people who perform it.

We use the term "control" in a wide sense. We include people factors like numbers, frequency and duration of the task, individual competency and fatigue. Arguably, these are all controlled implicitly or tacitly. We would also include environmental factors like temperature, lighting, noise and the nature of the physical environment (e.g. adequacy of flooring) to be included in our concept of controls. All such things are under the control of the designers of the workplace, the users, maintenance functions, etc.

There are some assessment models that promote a double assessment process – the risk before controls are considered and again with controls in place. This is possibly due to confusion as to whether the control is of the hazard or the workers' exposure to it (i.e. controlling the likelihood of an incident occurring). Anyway, is there really value in making this additional effort? Exactly what constitutes "no controls?" Let's take the example of a welder. No controls would be someone who couldn't weld, sitting in the nude with welder in hand. OK, a daft picture to paint perhaps; so the person has the competency to weld and they have overalls, boots, gloves and a mask, but aren't they the controls that they'd be supposed to have? The model that we promote in Chapter 5 will take into account all of the controls that presently exist, or are supposed to exist, but might be missing or not working. This way we can reach an accurate picture of precisely what is going on and, importantly, whether we need to act further.

For many activities, the controls will have either been dictated centrally or be an industry standard (such as the welding example), hence the central task-based risk assessment templates are useful in terms of noting which controls should be in place as a standard. This of course may be a surprise to some managers who didn't know that certain controls were organisational policy.

Pitfall 9: Whenever we review a risk assessment, we must add more controls

We have made some comment about continual improvement in previous chapters. It is a common belief of many safety professionals that you have to show continuous improvement. This is not the case. HSWA74 in the UK dictates that we need only control risk as far as is reasonably practicable. For risks that are already well controlled, although additional controls may be possible, the cost of implementation (time, effort and finance) may be disproportionate to the benefit that would accrue. In such cases, efforts should be switched to ensuring the maintenance of controls rather than the application of new ones.

Summary

- Organisations must ensure that they have commonly used definitions for safety terminology and a strong discipline for using them correctly.

- Risk assessments are management tools rather than vehicles for compliance. They provide evidence for whether the existing controls are applied well enough or whether more effort is reasonably practicable.

- It is essential to clearly communicate the context of the task so that the risk assessment is focused properly and managers are appropriately informed.

- For large or complex organisations, template risk assessments that are completed locally are a useful replacement for the true generic risk assessment.

- A risk assessment is inadequate if it does not contain enough information to be informative to the manager as a working description of who, what, why, where, when and how.

- Rating hazards is the first step in the prioritisation of risk, the precursor of effective risk management. (Rating likelihood as well enables calculated risk levels and prioritisation.)

- Using lost time experience or prediction in hazard rating confuses the use of any hazard scale.

- Reviewing risk assessments does not mean that more controls have to be implemented: continuous improvement is not a requirement of UK law.

5

Risk assessment tricks and techniques

Building on the ideas introduced in Chapter 4, the description of the risk-led approach moves on to describe a simple qualitative risk assessment technique. It details the importance of setting context before hazards are identified and described or likelihood evaluated. A description of hazard consequence and details of what it is may be included when determining likelihood is then explained.

For simplicity, the book describes a 3 × 3 matrix as an example of how hazard and likelihood combine to provide a description and rating for risk. It is made clear what a risk rating means and how it should be interpreted as a label of relative significance.

There are some introductory notes on making recommendations for action based on the results recorded in a risk assessment. The chapter concludes with some considerations of how to design a risk assessment document.

We have discussed at some length that risk assessments are not just compliance documents but are also important information sources for managers. We recognise the legal requirement in the UK to have a formal written assessment in the workplace (where we deem necessary of course), but what we haven't discussed is what to include in a risk assessment that will make it a useful management tool.

In this chapter, we provide guidance on the recording of general risk assessments:

- Discuss how to add context to further embellish the task description in a synopsis.
- Explain how to carry out a hazard and likelihood analysis and rate them to label their relative significance.
- Discuss some of the key administrative points usually contained in a risk assessment.

In this chapter, it is assumed that the assessor is seeking to write a template risk assessment that will then be completed by local staff. This is the approach we favour for large or complex organisations. Of course, small organisations will not need to write a template; involving local employees and managers will ensure that the right information is recorded when the document is written. In large, complex organisations, a template risk assessment containing the common findings and listing the normal controls required is a helpful start to local teams and helps to ensure a consistency of quality across the organisation. It should still be completed locally to ensure that locally relevant, specific and uncommon information is recorded, especially if it may change risk calculations.

Over the years, some have advocated that risk assessments should be signed off and sent out from a central point. Others have favoured issuing blank sheets of paper and demanding the creation of all risk assessments from scratch in each location. It has been our experience that the practical solution may lie somewhere between the two approaches. Central risk assessments tend not to gain the local ownership that is desirable. As for locally produced assessments, they seldom actually happen. When they do, it is difficult to control the quality, content or style.

First set the context

In Chapter 4, we heralded our view that a risk assessment could have its task title supported by a synopsis – a section in the document that further describes the context for task-based risk assessment.

There are no rigid rules for writing a synopsis, anything can be recorded if it adds to the comprehensive nature of the document and provides relevant information for the manager. Coincidentally, if the synopsis is written to be primarily of use to the organisation's managers, it will also provide quality information to those external to the organisation that might have cause to challenge it, such as enforcement officers, lawyers and auditors. A synopsis may be a few sentences or paragraphs and rarely any longer than a page. If the synopsis is lengthy, it is likely that the task title is too general and needs to be broken down into subtasks.

The synopsis provides more contextual information and may include an explanation of the following:

- Why the task is operationally significant?
- What is involved in completing the task?
- The industry perspective, i.e. reflection on accepted or best practice.
- Relevant government or industry guidance and why it is adopted or not.
- Who may be exposed to the hazards?
- Any appropriate organisational policy that may apply.

There are some very helpful words and phrases that can be included in the synopsis that make it more flexible than it otherwise would be, such as "typically", "generally speaking", "in the majority of cases" and "as a rule". There has to be the opportunity for local assessors to disagree, question and add more detail to the synopsis if needed. Therefore, the synopsis merely provides a template so that everyone agrees on what it is that they are assessing.

EXAMPLE GENERAL SYNOPSIS FOR A MULTI-SITE ORGANISATION WHERE THE TASK IS REPEATED IN DIFFERENT LOCATIONS

Task: General access to the car park and outside areas
   Synopsis:
Employees walk around the site during their work. In addition, customers, contractors and other visitors may use the car parks and wish to access the shops or adjacent offices. As a rule, this organisation seeks to use painted walkways, pavements, zebra crossings, lighting, fencing, and vegetation to channel pedestrians to walk in areas away from vehicular traffic at all of its sites, thereby assuring segregation. Such controls are implemented as is appropriate for the site but generally modified to account for local conditions. In so doing, ABC Ltd. endeavours to mirror the standards of the public highway using the same signs and road markings where appropriate.

   In the design and building of new car parks and site access, architectural plans are reviewed to ensure that vehicular and pedestrian movement have been segregated wherever reasonable to do so. This aim is also reflected in the company's construction design standards, maintenance contracts and general site inspection and maintenance arrangements.

Rubbish accumulation, broken slabs, kerbs and uneven and unexpected level changes can cause trips. Slips may be caused by wet pavement and algal growth; therefore, the maintenance and cleanliness of these areas is also taken seriously. In winter conditions, ABC Ltd. follows the guidance published by the government by favouring the gritting of key turning points and bends on-site roadways and by clearing snow and salting pavements where there is significant pedestrian traffic. In autumn conditions, similar arrangements are made for clearing away fallen leaves.

Some pedestrians visiting the site are disabled or blind. ABC Ltd. ensures that the design standard it adopts makes provision for this special group where it is reasonable to do so.

Why is a risk assessment needed at all?

Before we explain how we would recommend hazard and likelihood analyses are completed, we reiterate that, despite what many people may think who are trapped in the compliance-driven model discussed in Chapter 1 (Figure 1.1), risk assessment is not the goal. While it is accepted that in many countries a risk assessment must be completed and significant findings recorded, it should offer more than just being a record that is filed away and only retrieved when the auditor or enforcement officer calls. A risk assessment is of no use unless it reaches a conclusion; either we need to do more or conclude that no further action to reduce risk is required. Choosing not to act to reduce risk is perfectly acceptable if evidence proves that the likelihood of accidents occurring has been reduced as far as is reasonably practicable. Even if they have not completed risk assessments before, it is highly unlikely that organisations will not have some form of control in place.

Hazard analysis

A hazard is something with the potential to cause harm. It's a simple definition and one that is widely accepted. The workplace is full of hazards, but we don't constantly worry about all of them, all of the time – why is that? We have determined in Chapter 3 that "task" is the context for the hazard; so, in order to understand why the hazard could be a problem, it is necessary to use reasonable judgement to determine how the person doing the task could adversely interact with a hazard – what we'll call the reasonably foreseeable incident, "What could go wrong and why?" This really is the crux of this section of the assessment. It naturally leads the assessor to judge what degree of injury (consequence) may be expected, which then allows a low, medium or high rating to be assigned to hazard.

Reasonable judgement is used to predict the injury extent or hazard consequence. As we discussed in Chapter 2, reasonable judgement is founded on competence. The assessors must use their competence to reasonably foresee what the accident would be. They use their competence again to predict what the reasonably foreseeable worst-case injury (RFWCI) will be. Care is needed: the assessor should not identify the worst possible injury as just about any reasonably foreseeable accident could then be predicted to lead to death. Nor should the most common or best injury outcome be sought. That would lead to either a false sense of security or the claim that nothing will happen at all. For example, a warehouse employee using a fixed blade knife to open a carton could slip and stab himself in the chest and die. Even though it's possible, it would be a bizarre thing to predict. A competent prediction of the reasonably foreseeable accident would suggest that a slip might happen, resulting in a cut needing a stitch or two. Some predictions can be obvious to judge, e.g. cleaning gutters and losing the balance to fall from the roof of a three-storey building is predictably fatal. Surviving such a fall with minor injuries would be possible but practicably, a miracle.

Injuries can be grouped into the low, medium and high categorisation:

- Incidents that (at their reasonable worst) predictably result in injuries that are entirely treatable with minor first aid, i.e. they will heal without professional medical intervention (Low).
- Incidents that (at their reasonable worst) predictably result in injuries that are life threatening, life changing or give permanent disability, i.e. they don't ever really heal (High).
- Incidents that (at their reasonable worst) predictably result in injuries that are, in simple terms, neither of the above. They must require more than minor first aid and will heal in time, but clearly aren't as bad as being life threatening, being life changing or giving permanent disablement (Medium).

There may be some who see this as far too simple. This is for them to decide and explain why it needs to be more complicated than that. (We will argue in later chapters how three categories are sufficient to enable prioritisation.) Figure 5.1 shows this scale diagrammatically and indicates that numbers can be assigned to the low, medium and high categories – more on the use of numbers later.

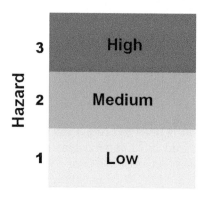

FIGURE 5.1
Hazard categories based on injury prediction.

Identifying likelihood factors

So, what is it the likelihood of? This is a crucial question. Some believe that it is the likelihood of the injury occurring, others hold the likelihood of the incident occurring. It is essential to remember that where likelihood is at its lowest, i.e. low, this is an acknowledgement that an incident could occur. Low likelihood does not mean none. We argue that in most organisations, the significant hazards and the incidents that can reasonably be predicted to result from them have been well known in the workplace for years. The question is whether the existing controls are suitable and sufficient or adequately implemented. We favour, therefore, the risk assessment approach that considers likelihood to be an analysis of what makes the predicted incident more or less likely to occur, i.e. given these controls how likely is the incident we predict to happen? We suggest that the cause of any incident during a given task is typically due to the failure or absence of one or more of the controls that should have been in place. Hindsight may suggest that a missing control would have been beneficial, but not always. In a mature system, it becomes less likely that more controls are needed in response to an incident; rather existing controls should be more effectively applied. This supports claims of being a learning organisation. If investigation indicates that a reasonably practicable control was missing, then a learning organisation would also wish to understand why it was missed.

Likelihood is about suggesting why these controls might fail or identifying any that are missing. Managers influence the adherence to controls. Some may be standard in the industry and others may be locally devised and reflect local custom and practice.

There is no set list of likelihood factors for every task-based risk assessment but in terms of providing a pick list of the most common factors that can be used to construct likelihood statements; these are some of the more obvious ones:

- Competency: one of the most vital requirements for any task being performed is that the operators are competent. Do they have the right knowledge, experience and ability?
- Physical engineering controls: guarding, trip systems, interlocks, etc.
- Safe systems of work, codes of practice, permits to work, etc.
- Signs and other information.
- Environment: lighting, noise, humidity, types of flooring, distractions, etc.
- Previous history of incidents: caution: absence of history doesn't mean that it won't happen, but frequent occurrences in the past may really be a strong indicator that it may happen again. More precisely, ask how many similar incidents have occurred over what timescale, i.e. three incidents in the last 10 years or during the last year, or month, or week or today will provide different urgency for control change.
- Questions that reveal context such as how many do this task, how often and for how long. We can ask about the frequency of the task but again care is needed. Does doing something more often have a positive or negative effect on likelihood? Does it develop experience or lead to complacency? Many hands may make the work easier or quicker, but too many people could lead to confusion, making the predicted incident more likely to occur.
- PPE, safety clothing and equipment.

If the organisation is a small enterprise on a single site, then the results of this analysis can be recorded as facts. For large or multi-site organisations, the organisation must allow the local management to record site-specific information. It is rare for a generic risk assessment to be equally relevant at each site. In such cases, a template risk assessment can be written by head office, but each likelihood factor needs to be phrased in the form of a question that can then be listed in the likelihood section of the risk assessment. The question is phrased to answer as a "yes" or a "no", but assessors should be encouraged to add in further information relevant to the site. The question should reveal the actions that should be taken, e.g.: "Do people routinely wear their PPE?" or "Are there occasions where people don't wear their PPE?" Either would be acceptable. Avoid making questions that are so absolute that they cannot be answered truthfully or without compromise: "Do all employees wear all of their PPE at all time?" Can we really say yes to this? Possibly a disciplined service such as the police could do so, but in a goods yard at a distribution centre or on a building site? The key is to be realistic with the question so that the assessor can be realistic with the answer.

The question for the assessor is "How good are we at all wearing our high-visibility jackets when we cross the goods yard?" Really good, really bad, or a bit hit and miss at times? The determination of likelihood is generally made "on balance" when reviewing the positives and negatives of the adherence to controls.

Determining the likelihood rating

If you don't like the likelihood rating and feel that it is somehow wrong, it may be that you simply have to accept the result as now you have better knowledge. Alternatively, you may have to re-examine each likelihood factor a little more and find where the result has been skewed. It can happen, but don't alter the result unless you justify the change with auditable or checkable data.

Assessing of likelihood cannot be done simply on the number of positive answers compared to the number of negative answers. It's simply the case that not all likelihood factors have equal weight. Which likelihood factors are most significant will be directly related to understanding the task in question and in what context the task is completed (e.g. with ample time to do it, or something we are not used to). Often a qualifying statement needs to be added to make the information make sense, e.g. six people doing this task 100 times a day could be interpreted in different ways because:

- They grow complacent and make mistakes.
- They generate a high degree of competence.

In terms of expressing likelihood, we again suggest a three-category scale with a low, medium and high response to the likelihood of the predicted incident identified and the data collected. Our three categories would read as follows:

- Unlikely that there will be a control failure leading to an incident. It is still possible to have an accident, but the factors that make an incident less likely to happen are in place and the task is well managed. On balance, the controls are in place and working well (Low).
- Our controls (or rather lack of them) are unlikely to prevent from an incident occurring. This is a task that we are not in control of at present. Or, it may be the case that despite our best efforts and intentions, there are no controls that we could implement that would improve the situation possibly due to overwhelming external influences, but it may well be a task that we are driven to continue doing

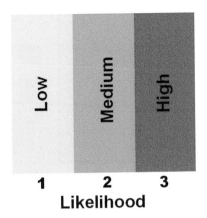

FIGURE 5.2
Likelihood rating.

for a greater benefit despite this realisation. Could a business operate an ice-skating rink without a pretty good chance of people slipping over? (High).

- Our controls are such that we cannot rule out the potential for failure leading to an incident at some stage; there is perhaps more that we could reasonably do to reduce the likelihood of an incident occurring, but the reality is that some controls are in place that are currently making a difference. The evaluation is simply that we cannot obviously rate likelihood as high or low. This result is an extremely common one and not an unexpected picture where controls were originally in place and over time have been allowed to degrade due to other systemic factors in the organisation.

Rating likelihood can be diagrammatically shown, as indicated in Figure 5.2. Once again numbers could be assigned to the categories as shown.

Assessing the level of risk

This part of the chapter should take little explaining. The truth is that you don't genuinely assess risk. You first assess the hazard outcome and then the likelihood of the predicted incident occurring. Risk is simply the multiplication of these two factors. The low, medium and high numbers in both categories are multiplied (see Figure 5.3). Multiplication follows mathematical convention since these are two independent variables. Bringing Figures 5.1 and 5.2 together provides a matrix with nine possible outcomes for the assessment.

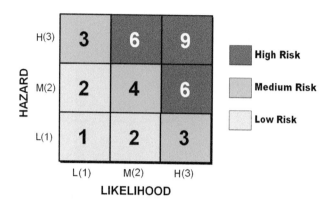

FIGURE 5.3
A 3 × 3 matrix for risk evaluation.

Refining the view on risk conclusions

At the beginning of the chapter, we outlined three basic outcomes that should be sought from a task-based risk assessment, i.e. "this is absolutely fine" and "this is not good at all" with the middle option of "this could be better, but it could be a deal worse too". Essentially, these are the basic outputs, but once you start to look at the way in which each outcome has been arrived at, you realise that there are far more conclusions sitting behind these three. We will explore these in more detail later on in the book when we come to making decisions in terms of risk management; for now, consider the question: "What is a high risk?" There are three options – three red boxes in Figure 5.3 that break down as follows:

- It's a task that we perform that our reasonable judgement indicates has a high potential either for controls to fail or that controls are absent that are necessary to control the likelihood of an incident occurring, which at its reasonable worst could lead to fatal or disabling injury (9).

- It's a task that we perform that our reasonable judgement indicates that the controls are fairly robust, but there is still a reasonable expectation for them to fail to prevent an incident from occurring which at its reasonable worst someone could suffer a fatal or disabling injury (top middle 6).

- It's a task that we perform that our reasonable judgement indicates has a high potential for the controls to fail to prevent an incident, which at its reasonable worst could lead to recoverable injuries following medical treatment and intervention (right-hand middle 6).

Obviously, there are similar permutations for medium and low risk. More of how you would make risk management decisions based on these risk levels is covered in the next chapter. Note that, when applying the concept of significance during task analysis to deselect tasks that are both insignificant and have insignificant risk associated with them (see Chapter 3), by default, all those tasks being assessed at this stage are significant. The numbers enable us to further define the degree of that significance. The numbers are nothing more than convenient labels showing the relative significance of these risks, but they do enable prioritisation of action, management and review. This is explained more in the next chapter.

Making recommendations

We began this chapter with the thought that risk assessments must end with a conclusion. It is that conclusion that must drive the next part of the process. It is important to remember that risk assessors, whoever they may be, should be the best people to assess the level of risk via the assessment process, but they may not necessarily be the best placed to recommend further actions should they be necessary. Clearly, we would advocate that they are nonetheless consulted with when the decisions on what action will be taken are made by the appropriate level of management. Any of the controls that were obviously missing, or were found to need more robust implementation, is a good place to start. There is no point looking for new, additional controls if you haven't been able to demonstrate that the existing ones are implemented properly. The evidence recorded in the risk assessment must drive the decision-making.

A good task-based risk assessment document must show clearly whether additional control actions are necessary or not. If actions are required, then managers should record who are responsible for actions and by when those actions must be concluded. Without names and dates, a recommendation list is nothing more than a wish list. With names and dates, it becomes a plan.

Other considerations for the document layout

There is little formal requirement for administration details on a risk assessment. However, the following are some of the considerations in terms of best practice:

- *Location*: this could be at a department level or simply the name of the site. It's what's needed to identify where the assessment is relevant.

- *Date*: having an assessment without the date that it was completed would be exceedingly unhelpful in many regards; however, consideration needs to be given to the date of when the risk assessment was completed locally (and if a large or complex organisation, when the template risk assessment was written too). Dates may also be accompanied with a version number that reflects that the risk assessment has had a number of different versions as knowledge developed.

- *Serial and version number*: for a large or complex organisation, being able to track documentation will undoubtedly be part of your existing internal processes. As indicated in Chapter 3, the serial number might indicate the tasks position on the task list. It can also have prefixes which identify the relevant business unit to which they relate, specific process areas or operational functions, e.g. A/M/MT/07 which may break down as follows:
  - A – division
  - M – maintenance
  - MT – miscellaneous tasks
  - 07 – the number for task title assigned on the task list
  - v1 – version number

- *Assessor details*: knowing who completed the assessment is useful. Being able to go back and ask additional questions or be able to demonstrate relevant knowledge of the assessor may well be useful in the future.

- *Review date*: there is a clear duty to review risk assessments and whichever interval or reason prompts review of your assessments, noting it on the document may be administratively helpful. It is worth considering that setting a review date, for the undisciplined organisation, might create a rod to be beaten with in the almost inevitable case that the date passes without a review. As an alternative, a "last reviewed" date might be used instead of using changes or incidents to prompt the necessary interval for that document.

- *Thinking has begun to change*: some organisations have taken the decision to not review any risk assessment rated as low, e.g. those that have been low for perhaps five years. It is unlikely in the extreme that a task rated as low risk would escalate unseen to be high risk. In effect, what these organisations are saying is that low has almost become the new insignificant.

- *Signatures*: Is it necessary to sign a risk assessment? There does seem to be a fascination with signing things, usually associated with some notion of acceptance of liability. The reality is that an assessor need not formally sign a risk assessment. Of course, being able to trace who wrote it is desirable, but a name is not the same as a signature.

It is not the responsibility of the assessor to have the assessment; it is the responsibility of the manager (on behalf of the organisation), so if anyone should be signing it, it should be the department manager.

What does "sign off" mean? It's certainly not about acceptance of responsibility for the activity and its associated level of risk. Responsibility is defined by the health & safety policy and is given to managers to discharge. That's what being a manager is about, and in UK law ignorance is no defence. Signatures can mean different things on a risk assessment depending on what you want one for. A signature usually shows that the assessment has been read carefully and determined as an accurate and faithful representation of local reality as far as present knowledge allows. It may show that it has been checked for quality, accuracy of language and correct maths. It could also be used for the responsible manager to indicate that they are happy with the level of risk associated with the activity and they do not intend to pursue any further reductions.

Having a quality checking process is very useful in noting the "obvious" mistakes that everyone else has overlooked. This function wouldn't carry additional responsibility; it's simply another useful stage in the process. Being able to demonstrate how the quality of information and the accuracy of the task-based risk assessment are assured is a powerful argument to add to any defence in court. More importantly, it means that it is more likely that the organisation has a good understanding of the risk and how to reduce the likelihood of incidents occurring.

Final notes

We have concentrated our focus on an example of a general risk assessment in this chapter. We recognise that there are other assessments that may require different approaches and techniques; however, our experience has proved that this approach works for the vast majority of situations. Where risks may be dynamic in the sense where hazards or conditions have an element of unpredictability, we have found that a predictive risk assessment will nonetheless reveal a suite of controls to apply. The knowledge of such controls must be part of the workers training and instruction so that they know what to do regardless of what they end up being faced with.

Summary

- A synopsis provides further information about the task and helps set a context for the risk assessment to follow.

- Risk assessment is not the goal; it is of no use unless it reaches a conclusion.
- A risk assessment records the evidence that dictates whether further action is required or not.
- Hazard consequence rating is based on a judgement of the RFWCI that would result from the predicted incident.
- Likelihood data must indicate whether the incident is more or less likely to occur and is focused on how robustly applied existing controls are and whether any controls are missing.
- A three-category decision for hazard and likelihood is a more intuitive process, i.e. it's obviously a high or obviously a low and if not one of them, then it's a medium by default.
- A three-category decision for hazard and likelihood enables a nine-box matrix enabling a means to label relative degrees of significance and enables prioritisation.
- Any action requirements recorded in conclusion should be accompanied with a "by who and by when" to make it a plan of action.
- An incident should promote a review, but it is not the case that it should automatically lead to the addition of more controls.
- Location, dates, serial numbers, version numbers, assessor details, review date and signatures are good aspects to be included on a risk assessment document.
- A manager is responsible for the risk regardless of whether he or she signs the document. A signature demonstrates that it is a true reflection of reality as understood at that time.
- All processes for carrying out risk assessment must include quality assurance.

6

Prioritisation provides freedom and clarity

DECISIONS.... PRIORITIES......

This is arguably the most important chapter in the book because it begins to clarify just what managers are responsible for. It discusses how risk prioritisation delivers informed decision-making and may be used to argue what can be done to control and manage risk. This includes differentiating between two risks that have the same risk rating using moral, resource or operational justifications.

The creation of a risk register is described and how the document helps management to plan a response to busy periods, delegate actions to others, communicate risk and generally target the limited resources of the organisation effectively against the risk profile the risk register describes.

This chapter discusses risk registers in their simplest form – a local list of tasks prioritised by risk level and recorded in priority order. Fellow practitioners and some enforcement officers have often challenged us as to why numbers should be assigned to task-based risks and the validity of such a valuation. In Chapter 5, we recommended the use of numbers as a means of labelling to show relative significance. Consider the alternative, not assigning a nominal value to task-related risk would mean that all issues would have equal priority and demand equal management. Managers intuitively know that this cannot be right; when questioned, they usually have a reasonable go at pointing out the "risks" that they think they should be most concerned about in the workplace. Unfortunately, reasonable guesswork may miss the important issue or inadvertently focus on the less important but most publically complained issue. This is not a firm basis for demonstrating quality judgement.

A good risk assessment method recording quality data found and recorded by a competent assessor will justify the relative significance of one task compared to another. Particularly, if the quality assurance of the risk assessment ensures that the methodology was followed accurately.

If a safety management system is compliance-driven as opposed to risk-led (as discussed in Chapter 1), then it is possible to be persuaded that all task-related risks have to be assessed and all require action now! That doesn't work in the real world. If you take that argument to its literal conclusion, then all task-based risks would have to be controlled at all costs and probably to the detriment of the organisation. This would definitely lead to needlessly "gold-plating" some risks and would drive the organisation to go beyond what is reasonably practicable. No organisation has limitless amounts of money or time. Fortunately, such a demand is not a requirement in UK legislation. That means choices can and should be made and that resources should be primarily focused on the most significant risks. That would be a reasonable approach and a sensible and proportionate response. The production of quality risk assessments provides the evidence for this decision-making.

Assigning number values to task-related risk as proposed in Chapter 5 enables a simple basis for prioritisation. In turn, prioritisation allows limited resources to be targeted to those things most deserving of action first. That is reasonably practicable decision-making in action; the numbers are also useful to justify degrees of action or even justify when no further action is needed at all, i.e. evidence-based decision-making. We have already argued in Chapter 3 that it is a misconception that organisations should strive for continual risk control improvement. This is counter to the legal test of reasonable practicability which suggests that organisations will get to a point where the costs of further action may be grossly disproportionate to the benefits accrued. In other words, the law does not require improvement at any cost. In effect, when organisations implement their risk control strategies, they are reducing the overall risk exposure on a reducing balance. There will come a point when all risk control strategies have been implemented and risk has been controlled as far as is reasonably practicable. The organisation must now shift the focus of risk management away from implementing new controls and towards the maintenance of those already implemented.

Creating a risk register

Having completed the risk assessments, the risk values enable the tasks to be grouped together into categories. This provides an initial prioritisation and indicates where the most effort is needed. If there are many similarly rated tasks to be managed, then it will be necessary to fine-tune the order of the list.

Figure 6.1 shows the basic theory, but we would stress that having assessed all of the significant tasks that expose people to significant risk, you prudently wouldn't worry about creating the risk register document until something had been done about the most significant risk – the "9". What does labelling a risk as a high-risk "9" mean? A task is being completed where employees/others are exposed to a hazard consequence of maiming or fatality, and when reasonable judgement is used to consider the likelihood of the predicted incident occurring, it is an event almost certain to occur at some stage due to the weakness of controls. Act first, the paperwork can follow!

So, how do you prioritise if two tasks have the same risk value? The first thing to note is where the ratings fall on the risk matrix (see Figure 6.2). A decision may need to be made as to which of the two task-related risks labelled A and B should be tackled first? Both are a medium risk of 3, so which one would you go with first?

The first thing to note is that you wouldn't be worrying about this question too much if you still had high task-related risks valued at 9 and 6s and medium risk 4s to act on first. We choose the medium 3s to make our point because the difference between them is more pronounced. One is high hazard and the other is high likelihood; so, which should be favoured? We would argue that it depends on a number of variables.

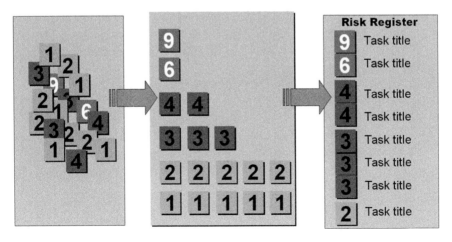

FIGURE 6.1
Risk prioritisation.

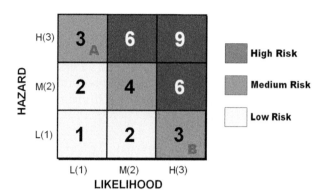

FIGURE 6.2
Risk choices.

There are many arguments for acting on one before the other:

- There is ample opportunity for a local manager to reduce likelihood by improving local controls, so risk B should be acted on first. They could ensure that workers are better informed, make changes to the workplace, write or amend procedures, provide better supervision at key points of the task, etc. This decision reflects a pragmatic view of risk management.

- Risk B involves minor injury predictions, but A has the potential to kill or maim; so, if we can do a little bit more to control it, we should. This decision reflects a moral view.

- For risk A, if the controls are not anchored, any degradation in their effectiveness can see the medium become a high risk; this may prompt further action to ensure that this does not happen by spending resources on the maintenance of the controls that are in place. It may also demand consideration of contingency planning, i.e. what do we do if the incident happens nonetheless?

- One of these risks may be linked to a task that is operationally critical. Work to identify additional control may even reveal a more efficient operational method of doing the task – a decision that reflects that safety risk is just one type of risk among many that an organisation needs to manage.

- Risk A also has the potential to cause significant reputational damage, so let's act on that first.

- Risk A (or B) has the potential to cause significant financial impact if the incident occurs, so let's do that one first.

- Risk B is causing complaint from employees, so let's do that one to help improve morale.

There are literally dozens of possible answers and all of them could be correct in terms of arguing reasonably practicable decision-making. In the UK, the application of HSWA74 allows business and operational considerations to be made when managing task-based safety risk, so these arguments are arguments of reasonable practicability that allow managers to manage according to local needs. In reality, of course, delegation will probably provide the opportunity to act on more than one task-based risk at once.

So now you have a risk register, how do you use it?

The first step is to double-check that the task-based risk assessments and the values assigned are valid, i.e. seek assurance that they are documents of quality. Without doing this, you may end up managing these task-based risks inaccurately, focusing your limited resources in the wrong places. So, how do you argue that they are quality documents? If you have followed the advice of previous chapters, you would probably argue:

- Involvement of more than one person to assess the risk, including the operators who are exposed to the risk and their safety representatives.
- Passing the risk assessment to the different stakeholders for checking.
- The inclusion of auditable data. My opinion versus your opinion must be avoided by the inclusion of checkable facts, measurements or other metrics whenever possible.
- That the risk assessment reflects the safety standards and recommendations for control made by central government or industrial sources. (Note: different controls may be implemented provided that similar safety performance can be demonstrated.)
- The scrutiny of the risk assessment by a safety risk management group or committee.

Once everyone is satisfied that the risk assessment is of quality, is suitable and sufficient and that in the risk register tasks are listed in the right priority order according to their risk, then it is time for action to be taken to control risk. Working down from the top of the risk register, the risk assessments should be analysed to see if any further risk controls are needed or, indeed, if it is already being managed effectively and adequately. A comments column on the risk register is useful for recording actions, e.g. the need for feasibility studies or the need for specialist advice, making additional risk control recommendations or any actions required to communicate or monitor risk. The risk register can then become a "one-stop" risk management summary tool. We would add that any action entry in a comments column should have the name of the person responsible for it and a date for completion too to make it a plan and not a wish list.

The types of tasks, the risk values associated and the number of them will all influence risk management action. Management may ask:

- Which of these task-related risks are intolerable and require immediate action?
- Which of these task-related risks would I wish to keep for my action and which would I be more willing to delegate? If action is to be delegated, can this be tied to the personal development plan for the person delegated to?
- Which of these task-related risks would be better dealt with by internal or external specialists?
- Which of these task-related risks could be tolerated for a while to allow our resources to be targeted elsewhere first? What kind of action or monitoring arrangements will be needed in the meantime?
- If resources are limited, what could I do this year and what must I leave to be included in my business plan next year?

The answer to all of these questions can be answered by using the risk register to guide management decision-making. Once a risk management plan is made, it needs to be communicated. In particular, employees who are at risk must be informed about it and how they are to be protected from it. Of course, most organisations would have been doing this already to some degree. No need to go over ground that is already covered by induction or task instructions, but you will need to consider if any new information changes previous instructions, procedures or training that needs to be communicated.

One other thing of note: if there are many risks to be communicated, perhaps it would not be wise to tell workers about them all at once, a false perception of danger could be inadvertently created. It is best to use the risk register to dictate sensibly what needs to be communicated, in what order and over what timescale.

The risk register can be used to inform local managers of the task-based risks they are responsible for managing. The risk register will inform them as to where they should focus their monitoring. It will also inform them which tasks they must more closely supervise to ensure that safety controls do not degrade, at least until they are completely confident that employees invariably apply the required control strategy. The risk register also informs local managers which tasks they should take greater care about explaining to new employees and supervise them doing in the first few days.

Reviewing the content of a task-based risk assessment is simple:

- Has the nature of the task actually changed the exposure to the hazard that demands the reasonably foreseeable accident and injury prediction need to be changed?

- Have changes in layout, of who does what or the locations where the tasks may be carried out, been made that are significant enough to change our assessment of the likelihood of incidents occurring?
- Did the findings of the incident investigation significantly change our view of previous history and therefore our grading of likelihood?

The position of the task on the risk register will also help decide how frequently the task-based risk assessment must be reviewed. Highs will require a more frequent review than mediums and certainly lows. For some high risks, it may be wise to review them before each occasion the task is carried out as part of a Permit to Work system. The frequency of review may also depend on the type of task being done as there may be an operational consideration to take into account, e.g. as part of a wider quality control process.

All these decisions are driven by the content of your risk register, the significance of the risks and the competency of those employees completing these tasks and the context of the organisation's operation. It is certainly a document that should be used regularly in management meetings to ensure that things are being managed effectively. Having task-related risk assessments, risk registers and risk management plans should be on the agenda for health & safety committees and be used to drive discussions. Since quality risk assessments record the evidence we have for the nature and extent of the hazard and the control of exposure to potential accidents, perhaps health & safety committees need to metamorphose into safety risk management committees.

Risk registers and task lists can be used when planning for a new project. For example, every year many retailers have to plan for Christmas trading. This is the number one season for trade. There is an increased throughput of product, more deliveries or increased numbers of items delivered to the store, less space in the warehouse, employees becoming tired by working overtime and so on. Such factors should be taken into account since they will probably change the assessment of likelihood on some risk assessments, i.e. predicted accidents can become more likely to occur. It may not be feasible to completely change all control strategies for every task-based risk in the branch, but by using the risk register, the planning team can incorporate actions to make sure that at least those risks at the top of the risk register are not compromised. Such a decision would be a reasonably practicable response to the situation and a demonstration of good risk management.

Safety risk registers can also be used to guide decision-making in times of operational emergency. For example, a distribution centre was effectively cut off by a major crash involving many vehicles closing the local motorway for over 20 hours. It created gridlock on all the surrounding roads. Deliveries could not get out and suppliers could not get in. Once the crash had been cleared, a greater than normal surge of delivery trucks arrived. Meanwhile, branches had been ringing up to demand deliveries as they were running out of stock. Operational pressure demanded that goods were processed quickly and loaded on delivery vehicles and sent on their way to waiting branches.

This led to an intensive and abnormal period of work. It was reasonable to foresee that employees and supervising managers would be tempted to cut corners. Management used the local risk register to identify which tasks and their associated controls must not be compromised and which ones could be relaxed out of operational necessity, a decision based on arguments of reasonable practicability. It had to be accompanied by very clear management supervision and communication. Of course, as soon as things returned to normal, the pre-existing control regime had to be re-implemented and supervised to ensure poor habits had not been generated – to do so was just good management.

In an organisation of significant size and complexity, whilst the concept of prioritisation through risk value holds true, having a single risk register is something that doesn't truly work. It is necessary to have subordinate registers, perhaps at a functional level like catering, maintenance, warehousing, transport and central offices. These functional risk registers will identify the specific tasks worthy of note. The causes of many of these issues within the functions may have commonality. In warehousing, the issue may be contact with moving forklift trucks and handling machinery. In catering, it may be being struck by reversing supplier vehicles delivering provisions. In central offices, it may be contact with vehicles in the car park. When the risk registers are merged, the details of the tasks become more indistinct and all of these functional perspectives on the risk may be listed together under a new title: pedestrian and vehicle segregation. The merger and renaming may prompt better project management and facilitate the bid for resources. As the registers are merged at different levels of the organisation from local to the board, there may be fewer issues on the register than the originals it was derived from. Of course, on this journey, this is to be expected as the registers pass through the hands of operational management who may have the authority and means to deal with the subject, without the need to pass it up to a higher level.

The main points covered in this chapter are as follows:

- Numbers assigned to levels of risk do nothing more than label relative significance.
- Numbers allow prioritisation.
- Prioritisation enables risk management, i.e. the targeting of limited resources.
- Prioritisation enables further arguments to support reasonably practicable decision-making (sensible and proportionate responses).
- A risk register is a useful management tool for informing management teams of what they need to take into account during the planning of projects.
- A risk register is a useful management guide when confronted with abnormal operational circumstances.

7

Finding solutions

In this chapter, we will discuss the effectiveness of each type of risk treatment. Having gone to the lengths of systematically identifying what could be the shortcomings of control for a given task, it makes sense that risk control decisions should meet that need.

The chapter includes a discussion of the hierarchy of risk control. It evaluates the effectiveness of engineering controls, safe systems of work, signs and alarms, ensuring competency and the provision of personal protective clothing and equipment. The points raised include an examination of how to make the decision about whether the provision of information is sufficient, when a procedural response would be better and when a safe system of work is necessary. The chapter concludes with further discussion on the concept of reasonable practicability.

Everything to this point has been about trying to identify what we do, why we do a task and how safely we do it. It may be the case that we are all completing our tasks very safely, probably because we have arrived at the existing control strategy through painful trial and error lessons rather than by being truly proactive when we first started planning for the task.

Nonetheless, maybe new controls aren't necessary; perhaps all we need to do is better implement existing controls.

Some UK safety regulations and guidance contain a control hierarchy. An employer can be requested to demonstrate the efforts that they have made in considering all of the available options starting at the top of that hierarchy. The choice very much depends on the task, the hazard and the operational context. Consequently, we will not provide specific advice in this chapter on what controls should be implemented for which tasks. We confine the discussion to the effectiveness of controls in general.

As discussed in Chapter 6, the control consideration for task-based risk pivots on the predicted incident. What makes this predicted incident more or less likely to occur? As we will discuss, controls can be task, hazard, hazard consequence or likelihood focused.

Basic control questions

In simple terms, there are three control areas to consider:

- *Preparatory controls*: How do we prepare so as to ensure that things go well (e.g. improving lighting where there is intricate work, provide training to improve competency)?
- *Execution controls*: What do we do when doing the task to ensure that things continue to go according to plan (e.g. isolating power supply, wearing protective gloves, interlock switching, following a safe system of work (SSOW))?
- *Recovery controls*: What do we do if despite our best efforts, things *still* go wrong (e.g. fire suppression systems, emergency evacuation procedures)?

You may recognise these questions and the control considerations they label. No risk control stratagem is ever finished until all three have at least been considered even if they are then discounted as irrelevant to the task.

Risk reduction

The inset shows the two most popular descriptions of a risk control hierarchy. Despite their general acceptance, we suggest that they are misleading. In reality, control discussions rarely relate to such a hierarchical and linear view, it is much more likely to be a dialogue that identifies a control strategy involving controls from more than one of these categories. Certainly, it is rare that one control will be sufficient.

RISK CONTROL HIERARCHY 1

- Elimination of the hazard or the task
- Reduction
  - Intervention controls, e.g. guarding, two-handed switches
  - Safety devices, e.g. fall netting, airbags
  - Safe systems of work (and supervision)
  - Signs/alarms
- Training, instruction and information
- Safety equipment, e.g. personal protective equipment (PPE), safety clothing, fall arrest harness

RISK CONTROL HIERARCHY 2

- Elimination of the hazard
- Reduce how many employees do this, how often, and for how long
- Isolate the hazard into a safe zone
- Control through organisational or technical controls, e.g. safe system of work, guarding, alarms

There are problems with both of these approaches. You only have to search the Internet to see that there are many different and often contradictory explanations for the categories in both of these hierarchies. Both are built on the suggestion that those control types at the bottom of the hierarchy are more likely to be weakened by error; therefore, it is argued that the control is consequently less effective. This is not a sound argument when error or deliberate action can negate isolation or engineering controls too. In reality, it is difficult to argue that signs and alarms are more effective than safety equipment since one may be ignored and the other not worn. Surely, the error argument belies the important and vital aspect that knits any control strategy together. Without a positive safety attitude in employees and good management supervision, any control can degrade to ineffectiveness. Good safety is just good management.

Business risk management provides a hierarchy of control strategy: elimination, transfer, reduction and toleration. These hierarchy lists in the inset both start with elimination and continue with descriptions of different reduction methods, but are arguably flawed because they do not include transfer and toleration. In safety terms, transfer can be made through engaging a competent contractor, while tolerance is recognising that either the risk is insignificant or that no further control is reasonably practicable. Nevertheless, we will focus on the elimination and reduction strategies in the rest of this chapter.

Elimination

Obviously, the most effective way to deal with a problem is to eliminate it. In terms of running a business, this may be difficult to achieve given that there will probably be a significant purpose attached to the undertaking in the first place. It's unusual to carry out a task that exposes people to a significant level of risk for no commercial benefit at all. Elimination is most simply achieved by stopping employees doing the task or removing the hazard. Some argue that it's possible to eliminate risk by preventing the projected incident from happening in some way and thus making it certain that the likelihood of the predicted incident is zero or at least negligible. To do so would mean that the employee would need to be kept from coming into contact with the hazard in some way. Fine, but if that is true, it is almost certain that cleaning and maintenance operatives would probably have to bypass that control. A different set of controls would certainly be needed for their tasks. In any case, what if that likelihood control failed? It may still be necessary to think about recovery controls.

General points about reduction

Managers often feel a sense of guilt if risk cannot be eliminated completely. They can mistakenly believe that to allow even a little "risk" to exist in terms of safety simply isn't allowed, which is nonsense and impossible too. There is no legal requirement to utterly eliminate anything with any level of risk attached to it. High-risk tasks can be performed as long as there is a justification and reasonably practicable controls have been implemented, e.g. it's all right to use explosives in a quarry, but we must exercise a high degree of control during the task and in the storage of the explosives themselves.

When looking at reductions in the level of risk, the same considerations can be used as we did for elimination by looking at task, hazard, hazard consequence and likelihood. Some of the most effective treatment is the reduction of the hazard or its consequence if it were realised. Less hazardous substitutes can change the nature of the hazard, e.g. changing a chemical cleaner to a less harmful one. Some kinds of safety equipment or clothing can reduce the level of consequence should the predicted incident occur, e.g. wearing gloves in a freezer, reducing the height of a fall using a net, or dampening noise levels. Hazards very often are what they are because we need them. High voltage has to be high voltage, knives have to be sharp and lift trucks need long forks. The operational reality is if we have to do that task for our operation to work, then those are the hazards employees will be exposed to.

Reorganising the task may reduce the likelihood of an incident occurring, e.g. placing safety-critical activity at the start of the task when people are fresh rather than at the end when tired. Performing the task less frequently or using fewer people can be a double-edged sword:

- Fewer people doing the task may improve the competency of the few as they become more experienced and able, but they may become fatigued or even complacent.
- More people doing the task may reduce fatigue and perhaps complacency, but competency can become dulled without practice; there may also be scope for them to get in each other's way or parts of the task being forgotten since it is assumed it is being done by someone else.

Reduction: engineering controls

We begin this section by questioning what we are seeking to control: the task, the hazard or the hazard consequence (injury extent). Are we trying to keep the employee away from the hazard, providing a fail-safe mechanism to provide protection if they inadvertently make contact with the hazard or are we trying to minimise the extent of their injury if an accident happens? We therefore question whether engineering controls is a description that is accurate enough? In our view, these controls fall into three categories:

- Prevention, e.g. machinery guarding, barriers.
- Intervention causing a fail-safe condition, e.g. light curtains, pressure mats, interlock switches.
- Hazard consequence reduction, e.g. car airbags, seat belts or side-impact bars.

The effect of user intervention affects engineering controls too. It's a common understanding that you must never remove, alter or bypass a guard (or other engineering solution) and to do so is often made a disciplinary offence.

Despite best efforts, operators still remove them and are injured. Some may even be tasked with doing so, e.g. maintenance engineers and cleaners who actively remove and bypass engineering controls as part of their duties.

Reduction: safe systems of work

The term safe system of work (SSOW) is often used imprecisely to describe procedure, process, instruction or simply the way of working (keeping the place tidy is not a SSOW; it's business as usual). We feel that in its purest sense, a SSOW is a written description of a routine that includes specific procedures that demand instruction (e.g. locking off, isolating and testing) and further hazard or operational information. Specific procedural detail need not be repeated in the SSOW itself, as that would create a cumbersome document. Instead, in the introduction to the SSOW, reference to the related procedures must be made with the caveat that no advance may be made unless these controls are implemented. It's following the SSOW that prevents the failure of other controls.

Not every task requires documents to help provide information, procedural instruction or SSOW. If they were required, it would be usual for them to be associated with a significant task and with a significant risk. Keep in mind that the purpose is to control exposure to a hazard in a situation where a degree of exposure is actually a requirement of the operation. It is the conclusion of the risk assessment that drives the need for these documents and the implementation of the actions they contain.

A SSOW is not always needed as we demonstrate in Figure 7.1. This model illustrates the interplay of complexity and effort in making a decision about what approach is needed. The decisions made in using the model need to

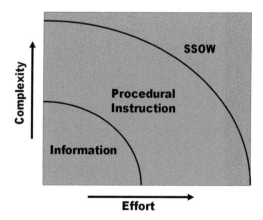

FIGURE 7.1
Model for choosing a method of directing action.

be context driven and reasonable as we discuss in Chapter 2; they may also reflect a consideration of the competency of those workers completing the task. By using the model in Figure 7.1, consideration of these two variables allows an X to be placed on the graph – where it sits indicates what approach is favoured. Note that the upper boundary line meets the end of the effort axis, but it does not meet the end of the complexity axis. This reflects our view that effort is not the principal driver for establishing a SSOW. Complexity, however, does drive the need for a SSOW. A very complex activity will need separate procedures, instructions and designated stages to maintain clarity and helps to guard against the failure of the controls it governs.

By complexity, we mean a task with many and varied parts and multiple procedures which must be performed in the right sequence in order to achieve the desired result. A task can be complex because of such things as the number of steps, the inter-dependability of different stages, how crucial quality is and competency needs. Using a ticket gun in a retail shop is not complex whereas completely rewiring an electric supply to a building may well be. A task can become more complex if it is a normal activity but performed in abnormal conditions, e.g. welding in a simple fabrication workshop compared to more complex welding when deep sea diving below an oilrig. Competency is relevant to complexity; the more complex a task, then often the greater the mental control, agility and skill needed.

We mean effort in Figure 7.1 in a broad sense but not in the sense of personal ability, i.e.

- how far likelihood factors are within the control of the organisation (see Chapter 5);
- how scientifically possible, realistic or even achievable other forms of control are; and the
- resources available, e.g. manpower, specialist availability, finance, time.

Since the first edition, this diagram has attracted a number of questions. In theory, most seem happy that the type of control is typically driven by the increase in complexity and effort; no one has suggested a formal SSOW is necessary for sweeping up the shop floor. However, it did generate some debate between the two authors, prompting this update to the diagram where risk is incorporated into the model as shown in Figure 7.2.

Figure 7.2 is intended to represent a concept rather than being prescriptive. Incorporating the organisation's risk matrix adds an additional layer of decision-making potential. In short, the decision as to which control to apply to a given task must surely in some way be driven by the level of associated risk. Clearly, the lowest risk can be served adequately by the introduction of information. The highest risk may be best controlled by a safe system of work, even a Permit to Work (PTW). Yet the boundaries between information, procedure

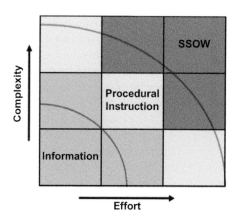

FIGURE 7.2
Risk and control choices.

and safe system of work cuts through different risk levels. This indicates that one size doesn't fit all. Sometimes low risk is best controlled by a procedure (e.g. the control of a slip or trip in an access route). Sometimes a high risk may not need a formal safe system of work (e.g. driving on company business).

This may be considered an overly simplistic model; nonetheless, we offer it as another idea to add to the mix when deciding what approach to take and which document to record. It illustrates that as things become more complex, there is a greater need to be procedural and to provide clear instruction about how to deal with the safety-critical parts of the task. High complexity and great effort may need a blend of procedures and therefore a SSOW approach. Considerable effort with low complexity may still only need procedural instruction to provide clarity and discipline. Low effort and low complexity will usually only require operational information to be provided.

We would add that procedural instruction might need elements of information; a SSOW may well require elements of procedural instruction and information. All may need training to provide the underpinning knowledge of why we do, what we do in the way that we do it. A SSOW and its elements may also need practice to develop skill. How many times have you heard someone in the kitchen say, "I don't understand it, I followed the recipe exactly?" Only for the reply to be: "you can't have or it wouldn't look like that would it?" In its simplest terms, a SSOW is just like a recipe explaining the necessary steps in terms of preparation, execution and, where necessary, recovery. It outlines the various types of equipment needed, the procedures involved and the sequence to be followed.

Both good SSOW and good instruction should only contain positive language and images. Taking the time to explain all of the things that must not be done really isn't helpful. It obfuscates causing a dilution of the message and creates confusion. There is a degree of skill required in writing these documents. Simply knowing the task inside out is not sufficient. There has to be an ability to write in positive terms and without ambiguity generating misunderstanding. Words such as *remember, take care, avoid, try not to* and *never* are all associated with a document that will certainly be negative. In preference, these documents should use words such as *must, shall, each time* and *always* (see inset example of good language use in an instruction that could be part of a safe system of work for cutting materials ready for fabrication).

The discussion is incomplete without considering the function of a PTW. It is not a control of the task. That is a function of the associated SSoW. A PTW is a formal acknowledgement that all of the necessary controls (contained in the safe system) are in place, have been checked, tested and it is now safe for work to proceed. PTWs are in many cases combined with the SSoW document (e.g. Hotwork Permits). This means that the difference in function between the two sections must be clearly understood and correctly applied.

EXAMPLE OF POSITIVE INSTRUCTION

Changing the band saw blade

1. Isolate the equipment using the padlock provided.
2. Test that the power is disconnected, by pressing the "on" switch.
3. Undo the wing nuts securing the cover and remove.
4. Retract and lock the blade tensioner with securing pin.
5. Unloop the old blade from the guides and tensioners and discard in the metal waste bin.
6. Replace the new blade and release tensioner.
7. Replace the blade guard tightening wing nuts by hand only.
8. Using top pulley, cycle the blade by hand to ensure that the tracking is correct with the blade running between the guide marks.
9. Re-energise power and return to use.

Reduction: signs and alarms

Signs are very often the first control that people consider and include statutory signs and local signs. They are undeniably a weak control owing to the following reasons:

- They become part of the "wallpaper" and employees don't look at them after a while.

- Those saying "Wet Paint" "Hot Surface" and "Steep Drop" have been the ruin of many and actually tempt people to do what they are there to prevent!
- They are often used as a temporary measure and then left up too long so that people no longer believe them.
- People of different languages may frequent the workplace and some may not understand written signs – the principal reason for European regulation in this area demanding design, pictorial and colouration standards.
- Too many signs are confusing to the employee and visitors alike.
- They are unhelpful in terms of defending liability – "We clearly knew the stairs were unsafe because we put a sign on them saying Danger, Unsafe".
- They are not a substitute for managers making sure that things are happening as desired.

As for alarms, there are many in most workplaces. They are ineffective if employees do not recognise what they are and do not know what action to take when they do hear them. In most workplaces when the fire alarm goes, someone has to shout for people to get out before they believe it's not a drill.

So, if the human element is so problematical, what is the most effective "people" control?

Reduction: competency

Competency is not an absolute, as we discussed previously in Chapter 2. There are degrees of competency ranging from the not yet competent, basic competency, technical competency and finally being an expert in something. Employees must have a clear understanding of the limits of their competency and know how to improve it. Where things generally go wrong is when there is an expectation, real or perceived that an employee needs to act outside of the limits of their competency. "Don't worry we all learn by our mistakes" are not words we like to hear coming from the dentist's surgery.

Competency is a blend of knowledge, experience and ability. Everything else such as training is merely a delivery method for one of the elements of competency, in this case, knowledge. A qualification in itself isn't competency; it's simply proof of acquired knowledge. Ability can be either physical skills such as being able to use tools or mental application such as being able to solve logic problems. Experience naturally can't be taught. Gaining new experience, in an environment where there is potential for significant mistakes to lead to adverse outcomes, must be done under supervision against a defined plan, i.e. an induction process. Testing people's mettle by asking them to do something outside of their competency is something done all too often to provide an opportunity for personal development. That's fine as long as getting it wrong doesn't lead to disaster.

So what competency standards are required? A quality task-based risk assessment will provide the steer for what knowledge and ability standards are demanded. Where appropriate, it may refer to qualifications but only in terms of setting a base line of competency for the role.

So, where does health & safety training fit in all this? UK Safety legislation regularly uses the phrase "training, instruction and information", however, little distinction between these three delivery methods is made. There appears to be a general consensus of opinion that all employees must have health & safety training. This usually consists of the ubiquitous slides covering the law, chemicals, PPE, risk assessment, electricity, ladders, etc., all to be taken at yearly intervals. It's blanket training in the hope that somehow the trainees will figure out what they need to know for the specific circumstances of their job. Such broad and imprecise training does little to reduce the likelihood of incidents occurring.

Being risk-led isn't simply "having some risk assessments" – it really does shape everything you do. Task-based risk assessments must be of sufficient quality to guide you to

- provide training when underpinning knowledge is necessary to develop knowledge and promote understanding of why we do these tasks in the way described in instruction, e.g. explaining the various hazards associated with moving roll cages full of goods around the shop. Training is also given to develop and maintain ability, e.g. police officers training in public order drills;
- provide procedural instruction in a simple and clear document as indicated by the task-based risk assessment, usually for tasks with significant risks, e.g. unblocking a jam in a rubbish compactor;
- provide information when an employee is demonstrably competent (e.g. give a wiring diagram to a qualified electrician) or the risk is relatively insignificant (e.g. how to change toner on the photocopier).

Reduction: personal protective equipment

PPE only protects the wearer and not others in the vicinity. Whether PPE reduces likelihood or not will depend on its type, use and management:

- Some PPE is worn to protect against a hazard of known concentration or force and therefore makes an assumption that exposure has already or is almost certain to take place. This type of PPE actually doesn't affect the likelihood of incident occurring – it doesn't stop the brick from falling from the scaffold. It may lessen the injury (hazard consequence) on the day, but it doesn't affect the estimation of the reasonably foreseeable worst-case injury (RFWCI) in the risk assessment. The helmet only reduces injury if the brick lands on it and if the person is wearing it. PPE like this cannot be seen as a risk reduction measure in its own right. It makes people safer, but the level of likelihood remains the same. At worst, this type of PPE can lull the user into a false sense of security.

- Other PPE may reduce the likelihood of an incident occurring. Providing a delivery driver with stout boots with good ankle support and tread will reduce the likelihood of a slip or a stumble.
- Wearing a high-visibility jacket makes you less *likely* to be struck by a vehicle, as long as the driver sees you. However, if you are struck, a thin brightly coloured jacket is not going to provide impact protection; the hazard consequence is the same.

One last thought about PPE. An employee working at height wears a lanyard to prevent them from reaching an edge. Clearly, likelihood is affected if the reasonably foreseeable incident is falling off, but does it eliminate the possibility of a fall? If the control assurance is such that employees don't always wear the lanyard, then we would suggest that likelihood probably couldn't be rated as low. The wearing of PPE by employees is strongly influenced by the local behaviour of peers, good leadership and sound supervisory management.

Reasonable practicability

A chapter on controls would be incomplete without some further discussion of reasonable practicability. It is often described as a balance between cost and risk and that's a fair summary, but in practice it has many more facets to it than that. Here we build on the points we have already raised in Chapter 2.

How far we go in risk reduction and managing the residual risk element have to take into account the fundamental question: Who are you and what are you trying to achieve? Reasonable practicability, therefore, can only be judged in the context of the size and complexity of the organisation and the kind of operation it runs. Large organisations can have a far more complex and varied risk profile, but some tasks are identical to those found in smaller organisations. In effect, this means that they may not have to engage higher standards for the task-based risks that similar but smaller organisations face, but they would be obliged to do more about the more complex and varied risk profile. Obviously, the large organisation will expend comparatively more resource because this risk exists in more locations, but the *solution* does not have to be any different. Practicability is not the test; *reasonable* practicability is.

Reasonableness exists on two levels: the reasonableness that comes with the competency of the individual manager when exercising his or her judgement and the reasonableness that exists within the organisation itself, i.e. organisational competency. As argued above, a manager's competency comes from an understanding of the significant risk in his or her area and from the knowledge and instructions that come from it. Organisational competency comes from a management process that includes a description of how managers make decisions, what the process is, what checks must be performed and who makes them?

The competency of the individual or track record of the organisation must play a large part in determining the ability to make sound judgements too. The common model in so many organisations is that if an employee is good at what they do, he or she is promoted to management, who thus brings to the posts a significant amount of knowledge. Managers don't need to be sent on blanket health & safety courses, managers need management training; it's that simple.

Reasonably practicable decision-making isn't just a defence to hide behind. It's the ultimate proclamation of a truly task-based safety risk management system founded on evidence. Good quality and accurate task-based risk assessments completed by competent people underpin and justify the decision-making. As discussed in Chapter 6, calculating risk ratings enable arguments of reasonable practicability. Without this, what is reasonably practicable cannot be argued with confidence. In such organisations, greater faith is placed in what external agencies and guidance say rather than what is internal reality, as discussed in Chapter 1. That's the path that leads to a compliance approach and is counter to true risk-led safety management.

In this chapter, we have discussed elimination and reduction at length, but what of the other two risk treatment strategies. We cannot round off the discussion of reasonable practicability without mentioning the two other choices:

- *Transfer*: as cavalier as this may sound, it's perfectly acceptable to give the problem to someone else. Of course, there are considerations to make in terms of liability, but others may bring more appropriate competency and equipment for the task. It's a practical solution: just consider how often your organisation uses a contractor.
- *Toleration*: in very simple terms, tolerate it for now until we have the means and requirement to change it. Tolerate is not quite the same as "accept" which conveys more of a "so be it" attitude. Sometimes the right answer is that enough is already in place, no need for any additional risk controls.

Our final note is to make it clear that if resources do not allow the implementation of a reasonably practicable control in the short term, then management will be expected to plan not only for when the control can be implemented but also what can be done in the meantime to manage this risk effectively with the resources that are available.

When is safe enough – safe enough?

"Safe" as a word means different things to different people. There are degrees of applying safety, from putting corks on the end of dining forks to simply not caring at all and making people to do hazardous jobs in sandals and shorts. In effect, safety has a scale from blatant disregard to the clearly mad or bonkers. So, how does an organisation find the right position on that continuum? When it does, do they stay there or does the position move?

It is madness for an organisation to choose a "safe enough" point below the country's legal requirement. In some countries, however, even the minimum legal standard may be an inadequate standard when viewed from the perspective of business need (e.g. shareholder desire, market positioning). Depending on the construct of local law, it can be very prescriptive, e.g. defined exposure limits to chemicals. Alternatively, like European law, it may be less defined terms and place the onus on the organisation to define it, e.g. the employment of legal terms like reasonable, sufficient, adequate or fit for purpose.

As Figure 7.3 illustrates, the problem with aiming to merely comply with the law is like trying to balance on a tightrope. Changes in operation may

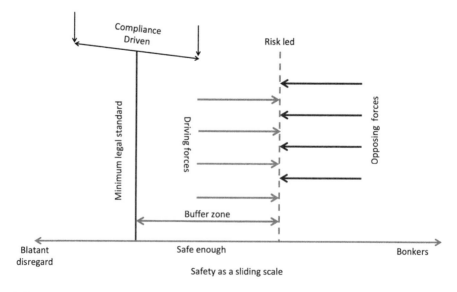

FIGURE 7.3
Safety is a sliding scale.

cause the organisation to wobble from side to side. If it falls the wrong way, then not only does an incident occur, but the organisation may also be judged to be out of compliance: a double hit. Focusing on mere legal compliance alone may generate a very fine line between safe and unsafe. This is uncomfortable as well as being time consuming and driving continuous uncertainty.

So, if we push the boundary a little and add more "safety", in effect we generate a buffer space. This means that although things may happen along the way as long as it's not catastrophic, the chances are that the organisation will not be proved to be non-compliant.

Implementing controls creates the buffer zone – not just any controls, but those that have been identified through quality systematic risk assessment and have been subsequently applied in a thoughtful and effective way. Caution is, however, needed. Simply adding more and more controls is not always a good thing. We argue that the concept of continuous improvement is not the most helpful idea. The suggestion that there is always more that can be done simply adds to the forces pushing towards the bonkers end of the scale in Figure 7.3. If we replace continuous with continual, it infers that whilst we should look for additional benefits, we need to only apply them if they make a material difference. As a wise man once said, "Just because you can, does it mean that you should?" Control levels can be increased to a point where they become so bureaucratic, resource hungry and complicated that they can no longer be effectively applied.

Controls need to be there only when they have a benefit. So, for example, a roofer wearing a high-visibility jacket is no safer than one who isn't. Who's ever been run over on a roof? Now if there are vehicles moving on the ground when the roofer comes down for tools and materials, then this is a reasonable control. But if it's just him and he has the only keys to the forklift, does wearing a high-visibility protection stop him running himself over?

More is not always better

Increasing the buffer zone by adding more and more control to push the risk-led line to the right of the model in Figure 7.3 can eventually adversely affect health & safety performance. Controls are, in essence, driving forces for better safety, but equally there are opposing forces. In order to move the line, it might be more efficient to remove some of the opposing forces, thereby decreasing the pressure pushing the line towards the left of the model. Reducing the resistance isn't always something that is obvious because of the way "safety" has always been seen as the addition of controls, especially following an incident.

Whilst not an absolute rule, the size of the buffer zone should be proportionate to the level or significance of the risk. Bomb disposal, for example, is going to want as large a buffer as possible where sweeping up the yard might just be a case of not doing it until today's delivery lorry has been and gone. Line managers must understand the concept of being safe enough. In reality, it is reasonable practicability rebranded. It's not about being safe; it's about being safe enough. Safe enough is all about the context of what we do and how we do it and proportionate response to risk.

An immediate criticism of this argument might be that safe enough sounds like a compromise. What would happen if we asked the employees working in an office environment if their workplace was completely safe? In our experience, people start to come up with highly improbable ways of being injured. There's something about the concept of being "safe" that appears absolute to people; if it's not absolutely safe, then does that mean it must be unsafe? Added to this, there is a feeling of personal liability. If I personally say it's safe (meaning completely safe), what happens if someone trips or pours coffee on themselves? Will I be liable because I said it was safe?

If we reverse the question somewhat and ask people if they think that on balance the office is an unsafe place to work, the likelihood is that there will be broad agreement that it's not and that there are no hidden horrors. In other words, whilst it might be possible to find something, it's really safe – enough. Safe enough gives sufficient latitude to be able to make mature decisions.

Summary

- In reality, risk control does not follow a linear hierarchy but is a sensible combination of different control types that satisfies arguments of reasonable practicability.
- Time is an important element in a control strategy. The strategy must consider what the ultimate goal is and include in the meantime controls.
- Good risk control is good management; without it, all controls will degrade in effectiveness.
- Maintaining a positive safety culture will help to keep risk controls effective.
- If a task or hazard cannot be eliminated, then by default the risk can only be reduced by controlling the likelihood of incidents occurring.

- Considerations of complexity, effort and the level of associated risk will assist in the decision of whether providing information, procedural instruction or a SSOW is required.
- SSOW includes procedural instructions that in turn are most effective when written in clear, simple and positive language.
- Permits to work are a tool to be used to formally acknowledge that controls are in place and working correctly; on their own, they do not prevent incidents.
- Competency (knowledge, experience and ability) is the key to developing a positive safety control and is the foundation of reasonable judgement but must be underpinned by an understanding of limitations.
- Accurate task-based risk assessments completed by competent people provides the evidence that justifies the judgement of what is reasonable practicability.
- Risk control decisions must be proportionate to the risk level and made in the context of the organisation's operation and risk appetite: to be safe enough.

8

Culture, involvement, behaviour and values

In this chapter, we will explore why it is necessary to understand why people do the things they do, what influences them and how you can affect the way that they see your version of the world. We discuss how an evidence-based, task-led safety risk management system managed by competent managers pulls everything into shape. To do so, it needs the acceptance of common values.

The chapter includes a discussion about culture. It examines why workers don't follow the rules. Why values for health & safety can and must support good decision-making by operational line management. How health & safety values properly set and applied will manage activity when the manager is not present.

We are not psychologists, nor do we claim to be experts in behavioural science, but we have learned some interesting lessons over the years from the school of life that we feel merit further consideration. The majority of these observations and experiences have simply been examples of good and not so good management. Poor safety performance is often attributable to managers who lack the basic ability to manage well and provide leadership.

The ability of managers to manage or lead is a matter for a different book. In this chapter, we offer some observations about culture, human behaviour and how managers can involve employees to help make the changes that you need. We will discuss culture, behaviour and values. It is partly about management systems, but it is also about management skill and ability.

Culture

For most people, culture seems to be an ethereal concept not easily described let alone measured. Uttal[1] describes culture as follows:

> Safety culture has been defined as consisting of shared values (what is important) and beliefs (how things work) that interact with an organisation's structure and control systems to produce behavioural norms (the way we do things around here).

Culture is a difficult thing to measure, but that has not stopped the demand for tools that can do so. Safety climate survey tools have been designed as a result and several are available on the market. The results that come out of using a safety climate survey tool indicate opinion, thoughts and feelings of how things are going. What is felt is usually nothing more than an indication of how good managers and employees are at communicating with one another. A typical climate tool claims to provide the following benefits:

- *Raising the profile of health & safety*: it gets people talking.
- *Monitoring*: it tests how well things are going before incidents occur.
- *Benchmarking*: it compares between locations or similar businesses.
- *Setting agendas*: it provides fresh inputs to the health & safety committee agenda.
- *Capturing sensitive information*: it allows anonymous contributions.
- *Working together*: it provides an impetus for manager and employee to work together.
- *Providing baseline measures*: an ability to compare results with the last one
- *Complementing audits*: it provides information that informs the changing of audit content.

Would a safety climate tool be helpful to an organisation implementing a task-based, risk-led approach to safety? We suggest that its value would be limited. The risk-led and evidence-based system we promote in this book delivers advantages in a different way through the following:

- *Raising the profile of health & safety*: involving employees in the identification of tasks, the completion of risk assessments and the development of controls will raise profile and provide much to discuss.
- *Active monitoring*: the advice in this book is to consider the likelihood of incidents occurring as part of task-based risk assessments. Being specifically related to task, there is no better basis for a good dialogue between manager and employee.
- *Benchmarking*: care must be taken since the same task at different sites may be correctly assessed at different risk levels; however, quality risk assessments should have sufficient information to explain why.
- *Setting agendas and working together*: What could be better for driving the agenda of a health & safety committee than a review of the task list, discussions around the quality of risk assessments, reviews of the effectiveness of controls and discussion about the implementation of any future risk control plans?
- *Capturing sensitive information*: seeking anonymous comment seems a little odd if the organisation is well led and managed, then communication should be open. If it isn't, that's a wider managerial problem and not specifically a safety issue.
- *Providing baseline measures*: we would prefer measures as follows: Does the site have all the task-based risk assessments it should have? Has it identified those stakeholders who should be involved in completing the risk assessment? Have local management and employee meetings included the discussion of controls for significant task-based risk?

Organisations often talk about changing culture, and it's not uncommon for them to look around for a prompt from outside their organisation to drive change. It may even be seen as an initiative: "Oh we did slips and trips this year, next year we have funding for a behavioural change programme…" In our view, this isn't how it works; you don't set out to change culture directly, it's what results from other more tangible changes that are made.

CASE STUDY 8.1

A company provided a brand new rest cabin for the shop floor employees to take their breaks in. Before long, sanitation problems arose with the floor becoming littered with tea bags. The wall near the bin was splattered and stained with the impact of refuse thrown across the room.

On several occasions, the room was cleaned up and redecorated, but the same situation arose. Appeals to employees to keep it clean and tidy failed to have an effect until finally the manager, in secret, painted the inside of the cabin in a vivid purple colour. This got the attention of the workers pretty quickly. When they demanded an explanation, the manager explained that this would hide the tea stains and be easier to keep looking clean.

This so shocked the team that they enquired what would have to be done to restore a more calming décor in which to take their breaks. Clearly, the answer was to keep it tidy. They redecorated the cabin in their own time with paint provided by the manager. This time, there was a feeling of ownership and a sense of pride in the job. You can imagine the reaction following the first person to drop a tea bag on the floor – "That is NOT what we do around here!"

Why do people do what they do?

In recent times, the subject of behavioural safety has been very popular. What is behavioural safety; is it as simple as people behaving safely? Does good behavioural safety provide a positive safety culture? What does a positive safety culture look like anyway?

If an organisation has followed the advice contained in previous chapters, then it will have systematic risk identification through task analysis, clear risk control action plans and assured competency. If you add to this list effective monitoring, maintenance, supervision and reliable incident capture, the organisation will be in excellent shape. Once an organisation has reached this point, there may be very little need for any further change and little point in discussing behavioural safety or further developing culture at all. Having said this, there are certainly some tips and tricks that may be useful.

For many managers, "why do people do what they do?" is a question that really gets to the heart of the matter. Typical answers range from laziness, taking shortcuts, not following rules, complacency and general self-interest. It's not unusual that the list is always about the employees and seldom about themselves as managers and their performance. Anyway, the fuller list is as follows:

- Because it's what they were *told* to do (hopefully correctly).
- Because it's what they *think* they were told to do (misunderstanding).
- Because it's what *they thought* they should have been told to do (initiative).
- Because it's what you reward them for (outcome over method).
- Because it's what they *have to do* to get it done (we haven't provided them with the right resources).

- Because it's what they are *allowed* to get away with (turning a blind eye).
- Because it's what they *see* others doing (peer pressure / learn by example).
- Because it's what they *want* to do instead (self-reward / shortcut).
- Because it's what happens when they make a *mistake* (error).
- Because they know it's the precise *opposite* of what they were told to do (violation/sabotage).

Why is it important to understand why people do what they do in the first place? If we are to manage them, we need to predict how they might misinterpret our instructions. Remember an employee's failure is potentially a manager's liability. We are not suggesting that each and every time a manager asks for something to be done, he or she must give specific, measureable, achievable, realistic and time-bound instructions, but it's worth thinking about for those activities that have a significant hazard consequence attached to them. Managers often give an instruction and are perplexed at the result, usually culminating in the usual exclamation: "If you want a job doing properly, do it yourself". Perhaps, they need to be clearer in their instructions and take the time to test the understanding with the employee so that they can actually achieve the result desired. Aren't those the skills of a good manager?

Somewhere down the "why do people do what they do" list, there is a change between your influence and their desire. That change comes very close to the bottom of the list. So why do people do what they do? Clearly, it has more to do with a manager than it does with an employee. So whose behaviour is really under scrutiny here?

If culture is "the way we do things around here", how are the ways that we do things defined in your organisation? Is there a culture of "do as I say, not as I do" or "I lead by example"?

Psychology 101

Much has been written on the subject of behavioural drivers and rather than recreate that here, we'd strongly recommend that you consider reading books from our friend and colleague, Professor Tim Marsh. Understanding the basics of psychology is the key to unlocking the secrets of human behaviour; and it's not difficult.

Why do people take shortcuts? It's simple; because they can and they see a benefit in doing so. That's not so complicated, is it? To remedy this, either remove the ability to take the shortcut, eliminate the desired benefit or create a stronger desire for something else. The problem is that workers seldom see the pitfalls because most of the time they get away with it unharmed. What's essential in a risk-driven and evidence-led approach is that we spot the shortcuts before it's too late. A point we have made throughout the book is that this is a core aspect of risk assessment.

Why do people fall off stepladders? That may be because they overreach. Why do they overreach? Because it's easier and quicker in their minds to lean rather than climb down and move the ladder. Once we acknowledge that this tendency will always be the case, we are better able to move forward to a solution. Simply telling people not to overreach will not work.

Nudge theory

People generally don't like to be told what to do, even when it's the right thing. Oddly, it can make them inclined not to do what is asked. "Don't forget to put some fuel in the car before it runs out" often leads to a game of how far can be driven and how many unsuitable fuel stations can be passed while the warning light is shown. We defy any driver to truthfully suggest they have never done this.

The trick is to be casual about these nudges where possible, building them into work routines and conversations so that they become part of that all so important culture that we're discussing here. Instead of saying "don't forget to wear your protective gloves", a nudge approach many be phrased as "let's all check one another's gloves to ensure they are serviceable". This generates a healthy concern to monitor the safety of colleagues, a reminder of the importance of wearing them and that safety gloves must be exchanged for a new pair if they are unserviceable.

A vacuum is easily filled

When a new employee joins your organisation, what sort of induction do they get? Be really honest here. In terms of health & safety, do they get an old training film to watch covering fire and manual handling? The trainer sets it running and then leaves the room asking them to switch off the terminal when they've finished. The trainer is busy that morning and has seen the film hundreds of times: "It's pretty boring, but it's a legal requirement". Such a scenario perhaps has recognisable elements – what impression does that give to the new employee?

CASE STUDY 8.2

A foundry business in the Midlands was flourishing during the late 1980s and as part of its expansion built a new casting facility. It needed to recruit a considerable number of new employees, mostly for unskilled and semi-skilled operations. The Safety Officer (clearly having more time on their hands than anyone important) was tasked with performing the inductions for all new employees. The Safety Officer decided that all new employees would start on a Monday at 08.00 hours and be seen all at once. There was a structured welcome to the business including its history, UK locations, its products, customers and an introduction to the management structure and key personnel. A pragmatic safety induction then took place covering what they needed to know, not simply a stroll through current UK legislation for no really good reason. The new recruits were then taken to the amenities block, issued with PPE, overalls, a locker and towels and given a tour of the site.

This took until around lunchtime. The employees were then split up into the various roles that they would be performing and placed in a specially designated observation area where they watched the jobs that they would be doing until it was the end of the shift. During this time, they were shown all of the activities that were going on and had all the things that should not be happening pointed out to them. It was only after all of this that they were asked to sign their training cards, once again reinforcing the comment that they were to do as they were told and not follow the poor examples that they may have observed. This was a well-established foundry with historical custom and practice that produced safety issues.

The result of this new approach was nothing short of miraculous. There were very few transgressions by the new employees, but any that were observed were corrected immediately. After a while, the new employees started to outnumber the "old guard" who were working in the new foundry section. Each new batch of workers slotted in with the existing crew and worked in the same correct manner. Eventually, which was never predicted, the new team started to influence the older workers. PPE use went up, accuracy of the work improved, problems due to not following procedures fell and retention was the best that the business had ever experienced. This may all sound almost unbelievable, but take it on trust, it actually happened.

Clearly, it is necessary to start as you mean to go on. If managers don't tell the new intake what they want to see, then who will? If a positive safety culture already exists in the organisation, then perhaps new employees could be turned loose for the workforce to educate them. If not, without supervision and support, the messages that are given to new employees will soon be knocked out of them by what they see and hear.

How to manage when you aren't there

It's neither possible nor necessary to micromanage everything and everyone all of the time. Exerting influence when managers are absent is about setting parameters within which people can make their own decisions. In the "why do people do what they do" list, we saw that for the main part it was about lack of understanding. If people clearly understood what is required of them and consistently acted as though a manager was looking over their shoulder, then there is a good chance that they will do the right thing. That requires very clear communication and the encouragement of self-discipline. Unfortunately, many of the messages that managers give people serve only to confuse, e.g.:

- "Don't do it if you don't think it's safe…"
- "If there are any dangerous bits to do, give me a shout and I'll come down and do them…"
- "Remember, if you think that I'd think it wasn't safe, then don't do it unless you think that it's safe…"

Managers mistakenly believe that such expressions convey how they want their employees to act in their absence. The information contained in task-based risk assessments and related work instructions must provide the necessary clarity to assist managers to explain their requirements to employees. Successful communication and understanding would help to ensure that employees carried out their work safely whether the manager is with them or not. Managers must ask themselves if they are happy that when away from the workplace (e.g. nightshift), their teams will be working in the same way that they would with them there. It's often the case that at weekends and on nights, there is a significantly reduced management presence. This often leads to a far greater degree of (enforced) autonomy where decisions are made and actions taken. Managers must ask if they want to know how they fixed *this* or worked around *that* or if they are simply happy with the outcome.

Values

In terms of performing actions, we have explored the role of the manager in influencing the right outcome. We must also consider why people are the *way* they are? We don't wish to open the debate of nature versus nurture here, that's a discussion for other books. At the risk of oversimplification, we believe it is mostly due to nurture, e.g.: Why do many people vote the way they do? Is it because they have read the various political party manifestos and been to meetings to confront politicians, or is it simply because that's the way their parents voted? What real influence does someone knocking on their door and handing them a leaflet really have?

It seems apparent that you cannot simply change behaviour by saying "change your behaviour", and therefore there must be some requirement to affect the programming that delivers that behaviour. In his book *Affective Safety Management*, Dr. Tim Marsh (2009)[2] discusses human behaviour in more scientific terms; our views are purely based on personal observation and not on scientific research, we therefore defer to him in the wider sense on this matter.

What we will say is that each individual has a set of values and beliefs, which fashion our behavioural responses. We may be largely unconscious of these values and beliefs as they are probably quite hidden: only becoming visible in our attitude or demeanour. These attitudes drive the way in which we respond in any given circumstance, which ultimately is the way we behave.

Let's take an individual with the value that dishonesty is wrong. One day he or she taps you on the shoulder and hand you back your wallet, still stuffed full of new notes after a visit to the cash dispenser. What could be more honest than that? That person could easily have walked off with that and no one would have known. Well, we hope that is the case. Perhaps there was someone else watching and this was a case of selective honesty. How many people believe that they are honest, until they are faced with a question of whether a dishonest act would benefit them and whether there would be any chance of ever being found out? Is this a case of "as long as no one suffers, then I guess it'd be alright?" So, imagine you are surprised to hear later that day that the person who handed back your wallet was caught stealing from the business.

The reality is that your value for honesty and that person's are not the same.

> Honesty is about not harming people, but a *business*, well, they can afford it. It's not like they don't know that it's going on; everyone is at it. If they didn't want us to take things they shouldn't make it so easy. Of course I'd never take from a person, which would be horrible.

So, as a manager we cannot simply imprint our values onto others by way of a statement such as "only do it if it's safe". What the recipient thinks safe looks like might not match your perception. What we need to do is have a set of ground rules that make it clear what you and your organisation think that "safe" looks like.

Drawing upon our experiences working with some of the emergency services, we drew up a set of core values that encapsulated our thoughts at the time. Whilst these values may well appear relatively universal, please don't simply copy them wholesale. Whilst you are welcome to use them, consider them and even dismiss them, we would not wish the reader to simply think that as they worked for us, they will work for you. As with everything in this book, we are putting forward a philosophy, not selling a management system to cure all ills.

We use three core values: nothing we do is worth being harmed for, unsafe behaviour will be challenged and not rewarded and everyone has the right to make a challenge and expect a response.

Nothing we do is worth being harmed for

There may be occupations where injury, whilst undesirable, is actually part of the calculated risk associated with that job, e.g. the armed forces, fire service, police or bomb disposal. Fortunately, the vast majority of organisations don't have such a risk profile. There are no tasks or parts of our undertaking where we wish employees to place their safety or that of others in second place. Everyone should go home unharmed and a little wealthier than when they arrived.

Work should not have an adverse effect on employees' health and well-being. This is a goal to which we should aspire, but remain realistic that this is a physical world and when employing large numbers of people, performing so many diverse operations, preventing all incidents may not be possible.

When talking about "harm", it is important to understand that it is significant harm that is the focus of this value. If we really meant any harm whatsoever, we would have to lock the doors and all go home.

Working hard and being tired at the end of each day is quite normal and is not what is meant by this value. Neither are the everyday bumps and knocks we all get as part of moving through life. The key message of which employees need to be really clear is that there are no circumstances in which they should feel it necessary to compromise their safety and well-being in order to get the job done. It is vital, therefore, that each organisation provides the necessary resources to be able to work in that way.

Unsafe behaviour will be challenged and not rewarded

This core value is about having a "just culture". This means that if the business has fallen short in preparing the individual, there can be no blame. So standing on a chair because there are no ladders is not the individual's fault. Let's not forget however that employees have a responsibility to tell us when they need something such as work equipment to do their job properly or that something that they were given is not working properly.

When instructions have been given correctly, the necessary equipment has been provided, competency is beyond question, and people still exhibit unsafe behaviour, perhaps it's because they can't be bothered to fetch the ladder; then it is only right that disciplinary action be taken.

Challenging behaviour needs first to be understood. We must not be quick to apportion blame or mete out punishment. It would be good to say that "safe behaviour will be rewarded". To be fair, all behaviour should be safe and of course it's your safety, so to reward you for not hurting yourself would be a little odd. Clearly, we want people to tell us when it's not safe and to bring ideas to us when they can see where improvements can be made. That is certainly something that we may wish to reward. For now, it is important that we project the message that we wish to find out what is behind any instance of unsafe behaviour so that we can do something to make it unnecessary in the future. Taking into consideration core value number one, it would not be acceptable to justify unsafe behaviour by saying it was done in order to get the job done.

Everyone has the right to make a challenge and expect a response

How many organisations truly have such an open and honest culture where any employee may challenge another, their manager or the business as a whole and receive a reply? This is not to say that everyone will always get the answer that they would like, but there is nothing worse than questions simply not being responded to which will eventually stop any questions or challenges being posed. Challenging unsafe behaviour is something that we agreed upon must be done to find out "why". This challenge is not just about managers challenging their teams; it's about teams challenging their managers too. If organisations wish employees to challenge, they should also encourage the challenger to have a suggestion ready or possibly explain what they are prepared to do to make things better too.

CASE STUDY 8.3

During a cold and snowy period of winter, a grocery home delivery company decided to give local managers the authority to decide whether deliveries should go out or not. This reflected the core value that nothing we do is worth being harmed for. The company had recently spent a fortune on educating its employees that customer service is of paramount importance. This led some managers and employees to believe that although employees should not place themselves in harm's way, if it was to serve customers then it was all right to do so.

One driver got his vehicle stuck in snowdrifts on a remote country lane. Refusing to be prevented from completing his mission, he took hold of the bags and marched a mile through blizzard conditions to make his delivery. Not only that but he made a second trip to and from the customer's house to finish the delivery.

The second core value requires that because he clearly placed himself in harm's way, the delivery driver should have been challenged and not rewarded. Instead, the company magazine heard of the story and wrote an article praising the delivery driver for his dedication to customer service. Even his manager was seen to publically thank him for his efforts in such adversity.

What kind of message did that send to other employees? How far would such a situation go to undermining the first two core values for safety? He should have been challenged to find out why he believed that this was what was expected of him. Next time he may not be so lucky...

Case study 8.3 demonstrates that core values for safety can be challenging to the operational circumstances of the organisation. That is as it should be. Safety considerations should be entwined with the decisions concerning how the organisation is run. This case study also illustrates the conflicting messages that employees can receive from managers.

As a final point on core values, take care to ensure that they are a value and not an objective. We have seen organisations claim that one of their core values is to have zero accidents. While we applaud the objective (which will never be achieved despite what people claim), it is not a value. The key is having values that lead to correct attitude and behaviour, but not only that, they must match the operation. Having the value about "nothing we do is worth being harmed for" simply wouldn't work for the armed services in defence of the realm. Clearly, we don't all fit into this category, but saying that we truly believe that nothing is worth being harmed for demands a very long hard look at the significant risk assessment findings. Doing so, we may realise that there most certainly are activities we'd rather not have that are central to our undertakings, but stopping them could actually limit effectiveness. Manual handling of fridge freezers into customer's property is clearly more than anyone would agree is acceptable, but to not do so would deprive the organisation of a profitable activity and customers of a service expectation. The real challenge surrounds the word "worth" in our primary value and if you can't sell that concept, then the reality is that you may need to aim lower – a problem for the board to discuss.

Building a perfect world

An important consideration to mention in this chapter is what is it that you want to achieve. We've been asked on numerous occasions to preside over culture change where it's clear that the client has no idea where they are nor where they wish to be –they simply want a better culture. Perfection is something that often stands in the way of good enough, but it is possible to define what an achievable perfect world might look like.

Consider this list below. Where would your organisation be if this was all to be true?

Wouldn't it be great if people:

- Followed the rules (methods, processes etc.)
- Said when they didn't understand
- Challenged others when they saw that it was wrong
- Stopped before it actually went wrong
- Spoke up when they saw situation changes that mattered
- Understood their own limitations
- Asked for help rather than struggled
- Gave help when asked; but offered before being asked

None of this is beyond being achievable, but in order to reach that position it is necessary to unpick each line and identify the factors necessary to make it a reality. Let's consider a few of them. If we want people to follow the "rules", what do we need? Well, some rules to begin with: the right ones for the significant risks. Not too many and not too few. Written in a way that's helpful and can be understood by all. Rules that reflect operational reality and can be put into practice. Rules that are kept up to date and reflect change.

If we want people to speak up when they see changes that matter, what needs to be in place? Firstly, there needs to be a common understanding of what matters. We wouldn't want workers running to managers every five minutes with trivia. There has to be a mechanism to be able to raise an issue, record it and feedback to the originator.

These are not major items, but it's worth looking at your own organisation and deciding if you like the outcomes in the list and just where it is in terms of making them a reality. Importantly, this includes implementation when you are not there. We recommend reflecting on the simplicity of this list and to envisage the quality culture that would result if your organisation could place a tick against each of these statements.

We finish this chapter by reiterating one key belief driven by our experience in the world of work – good safety is just good management. We summarise this whole chapter by saying again that the effectiveness of an organisation's safety system is most strongly influenced by the ability and skill of its managers and how well they communicate with employees.

In summary, the key points in this chapter are as follows:

- A positive safety culture flows naturally from an effectively implemented task-based risk-led management system.
- In the main, employee behaviour is governed by the ability of managers to manage and the quality of their communication.
- Task-based risk assessments and related instructions are vital sources of information for managers seeking to clearly communicate behavioural requirements to employees.
- Core values for safety not only drive the safety approach, but they can also be powerful enough to challenge organisational decision-making and any conflicting messages that are inadvertently given.
- Good safety is good management performed by motivated, skilled and able managers.

References

1. Uttal, B. (1983), The Corporate Culture Vultures, *Fortune Magazine*, 17 October, page 66.
2. Marsh, T. (2009), *Affective Safety Management*, London: IIRSM.

9

Reactive safety

Despite applying effort to identify, assess, control and manage risk, accidents can still happen. In this chapter, we will look at the reporting and recording of incident data and argue once again why language is so important in shaping a management system. Discussion is made about what to report and the form that those reports might take in order to achieve the appropriate response.

Problems with incident reporting are described, including examining the accident triangle myth that many still adopt. Some similarities between risk assessments and incident investigations are described and their relationship examined. An idea for deciding on the depth of investigation is discussed. Finally, some suggestions on what should be included in board reports are made.

Incident reporting and incident investigation are the "what and why" of the reactive part of the risk management process. In this chapter, we have not sought to suggest how incident reporting or investigation should be carried out, there are many sources of advice on this aspect. Instead, we examine some aspects of this process and try to shed a different light, perhaps to the extent of dispelling a couple of mistaken but commonly held beliefs in the process.

Before we begin this chapter, we would counsel against the reliance on incident data as the main driving force for change in any organisation. Just because an incident happened does not mean that it will happen again. Just because it didn't happen does not mean that it can't or won't. We would also counsel against comparing trends over a long timescale. Most organisations these days have to continually reinvent themselves to stay efficient and competitive. So, what would be the value of comparing last year's statistics with perhaps even three years ago when we were differently organised, had different competency standards, were of a different size serving different markets, etc.? Misplaced faith in statistical information coupled with a failure to understand the detail of individual incidents has caught many organisations out.

That being said, we are not saying that incident data is of no use; we merely wish to caution against using this most common of lagging indicators in what should be a proactive and leading safety management system based upon largely the predictive indicator that is risk assessment.

Incident reporting

We start with the problem with language. What is the difference in meaning of the terms incident, accident and near miss? (See inset for ours.) You might also pause to reflect for a moment on just how good your organisation is at reporting and recording accidents? It's a fair question to which the reply is often "Pretty good". "So how are you at near miss reporting?" This second question generally elicits a less confident response. Why is this? The fundamental problem here is the use of the words "accident" and "near miss" in that one traditionally infers injury and the other does not. So, if accidents are reported well but near miss stuff not so well, then the differentiator is actual injury. The reality is that the majority of organisations actually report *injuries* not accidents at all.

This idea needs further exploration; imagine an employee reported that:
"There has been an accident on the shop floor."
"What's happened?"
"Someone has broken their leg."

Not a particularly unusual story, but it's the response to the question, "What's happened?" that is particularly interesting. The answer received is frequently in terms of injury rather than details of the event. If managers hear of an accident happening where no one was hurt, the response is usually "Good, well keep me informed". Realistically, should managers be expected to drop what they're doing to investigate that someone fell over but was unharmed? It's doubtful. We are not necessarily suggesting that they should, but if some managers are not particularly motivated to investigate incidents with actual injury outcomes, it's easy to see why a near miss would be ignored. Managers need a technique to be able to filter which near miss is important and warrants investigation and which is not.

LANGUAGE: ACCIDENT, INCIDENT, NEAR MISS OR NEAR HIT?

The most interesting contrast for any organisation is the number of actual injuries compared to the reasonably foreseeable worst-case injury (RFWCI) that could have resulted from the incident. What we mean is that we must apply the near miss concept and capture that in reporting and investigation. If an employee walks between a reversing lorry and the dock, it's an obvious near miss/near hit but most organisations would rarely record it. If the employee was two steps behind and the rear of the lorry pinned him or her against the wall before it stopped, it would be a potential fatality even though the lucky employee may have got away with a cracked rib, but it would be an accident classed as a fracture. In this second case, an injury has been caused and would enter our incident statistics for fractures. We would argue that in each scenario above, the incident should still be treated with the same urgency and seriousness as if a fatality had occurred.

- We regard an accident to be a description of an event resulting in any injury.
- We regard a near miss to be a description of an event that could reasonably have resulted in any injury.
- We use the word incident as an encompassing term for both.

The First Aid Record Book that organisations keep can hamper record keeping of near misses. It's a record of how the injury was caused and the treatment given; so how often does anyone record an entry where no injury was suffered? It's unlikely that a First Aider would even have been called, so who is going to fill out a report of the event?

In the UK, the HSE's publication HSG 245[1] is another example of where confusion pervades – regardless of content, its title "Investigating Accidents and Incidents" just serves to add fuel to the language bonfire. What happened to near misses then? All of this flexibility of language tends to draw divisions in a management system where the opportunity for misunderstanding, dual standards and gaps can appear.

Linking incidents with assessments

With good risk assessments like the ones advocated in this book, there is an opportunity to tie the proactive and reactive sides of the safety culture closer together. Our view is that a risk assessment should identify what people will be doing, what could go wrong, why, how badly someone can be injured at its reasonably foreseeable worst and in reality how close to that position are we currently operating. In incident reporting, we'd consider what were people doing and what went wrong and how badly were people hurt? In investigation we add why did it go wrong, i.e. what were the likelihood (control) failures? We would argue this is a past tense mirror of predictive risk assessment and includes the examination of most of the same factors.

In a risk assessment, the thought of there being no injury does not occur – the assessment would have concluded that the risk was insignificant – yet people do walk away without a scratch following an incident. So, we advocate that which injury did occur should not be the main focus of the report, but that it should be what the reasonably foreseeable worst-case injury could have been. In other words, the technique for rating hazards discussed in Chapter 5 is just as applicable in incident reporting. This draws the attention back to understanding the event. It is a more consistent technique and is not open to the vagaries of luck.

We must have all heard expressions such as "He was lucky, 6 inches to the left and it would have hit him" or "If I had a pound for every time, I had nearly fallen in there..." None of this is helpful after the incident has occurred. In the system that we advocate, we record actual injury and the RFWCI in exactly the same way as we do in the risk assessment section denoting them as low, medium or high, or as level 1, 2 and 3 (see Figure 5.1). The level of RFWCI rather than the actual injury is then used to direct the level of investigation required.

We have discussed with peers whether this is something that we could consider as a leading rather than lagging indicator. Well, the incident will have still happened and some injury may have resulted, but as likely as not, the full and reasonable potential will not have been experienced. In that case, the difference between actual and virtual certainly could be considered as leading.

In a truly proactive, risk-led management system, we would argue that the leading indicators are the risk assessments. What better prediction of where you may be headed can you get than a risk assessment? The whole approach if done well will provide the leads you need to identify gaps and weaknesses in your controls and further lead you to identify if further controls are needed or whether existing ones need re-establishment.

There are a number of publications that advocate the concept of a "pre-mortem" where the task or project is deliberately subjected to ways to make it fail. We argue that this is the purpose of a risk assessment and really isn't anything new at all. In fact, we'd venture to say that by and large it's actually a less complete process than a well-conducted risk assessment.

What, then, is the definition of incident? In our view, it's simply where something undesirable happened, whether an injury was suffered or not. We certainly could not say that it was unplanned; you should have made a reasonable prediction that it would happen in a risk assessment remembering that likelihood even at its lowest indicates that failure is reasonable. So, unless it was a bizarre event, we did actually plan to avoid it by introducing preventative measures. The unstable load in the racking that with just one more nudge would have fallen is still down to a control that has been compromised and by now we should have understood how to deal with those situations and we should try to find out why the control failed.

Setting boundaries

In our experience, using the hazard rating scale to rate incident types works very well. If the effort has been made to ensure that the organisation understands the hazard scale, then incident rating using RFWCI is a natural overlay. The three-point scale we prefer lends itself to a straightforward reporting escalation system:

- Where the incident is a RFWCI low hazard or level 1, you might simply require the facts to be recorded. Conducting detailed investigations of accidents with a RFWCI of basic first aid might not be the best use of resources. You would not be preventing tomorrow's fatality.

- In the case of a RFWCI of medium hazard or level 2, it may be that it is the role of the local manager to conduct an investigation into the likelihood factors that led to the event occurring. They own those local controls and will need to understand what compromises might have occurred. If any failings in main management systems are identified, the findings will need to be communicated to a wider audience.

- Finally, a RFWCI that is a high hazard or level 3 incident should involve the senior management team and, when the organisation is large enough, specialists from facilities management, safety, occupational health, procurement, etc. Clearly, there are some significant lessons to be learned from this event. This is the level that should be reported to and discussed at board level.

You might argue differently to these action descriptions at each level and that might be right in the context of your operation. We have found them to work well in different organisations that we have worked for.

Reviewing risk assessments post incident

Linking risk assessments and incident reporting to examine likelihood controls in the way we suggest does make the comparison of the two very simple. What was the task that they were performing? Was this actually covered by a risk assessment and if not should it have been? Comparing reality with the projection that was made initially may reveal an uncanny accuracy. This should be the central point for the investigation. If an accurate prediction of what *might* happen was made, how was it *allowed* to happen? There are exceptions to this of course, i.e. where likelihood factors in the risk assessment predicted that there were elements that were simply beyond any control.

Comparing actual injury suffered with RFWCI may reveal that the full potential was either not realised or the injury experienced was not reasonable to predict. In the case of the latter, the RFWCI may need to be reviewed unless of course either the additional level of injury can be attributed to bizarre circumstances or that the increased level of injury suffered was due to a pre-existing condition in the individual that exacerbated the outcome.

Where this process really does start to add a new value is in highlighting trends that are not necessarily reflected in risk assessment as they are, technically speaking, tasks performed incorrectly. It's true to say that learning lessons retrospectively is not the best way, especially in what is claimed to be a truly proactive management environment, but what's the alternative – that you don't get to hear about these incidents?

CASE STUDY

We use portable stepladders for work. We've deemed that these are appropriate and accept that there is still a level of residual risk attached to using them. We acknowledge in our risk assessment that if workers do not move the ladders as the task laterally progresses, they may overreach and possibly fall and injure themselves. For this reason, our training focuses on the reasons why the steps must be incrementally moved. Observation indicates that workers comply with this requirement.

In a bid to ensure that the business remains competitive, there is a drive to deliver more cost-effective contracts and purchasing. Somewhere along the line, a new contract with a different supplier is agreed upon and whilst the ladders now supplied look the same, they are not. This has opened up a new hole in the cheese. Whilst the new equipment may not be intrinsically unsafe, it no longer fits the "rules" that were drawn up. This is not a reason for staying with the original supplier, but there must be a mechanism in place that receives sign off from critical stakeholders when anything with the potential for change is introduced. Depending on the context of the task and the change, this could be anything from new soap in the restrooms to changing the provider of company vehicles.

Reason's Swiss Cheese: model or theorem?

Some criticise the idea of Reason's[2] theory claiming it to be too linear. This may have merit, but the problem lies in the fact that it's not really a model for incident causation at all; it's more a concept that says for any given scenario, there are gates or opportunities to put in risk controls that make any path through the "cheese" less likely to occur (see Figure 9.1). This may be the competency of the individual or decisions made by a corporate board. The individual slices themselves should not necessarily be seen as being stacked as they appear in the popular diagrams where holes line up, physically. That is too literal in reality and sometimes the open route through the cheese is more convoluted than the simple illustration suggests.

Reason's theory is an excellent vehicle for educating managers on how incidents are allowed to occur when it was believed that everything was under control. No matter how well something was controlled, the inevitable truth

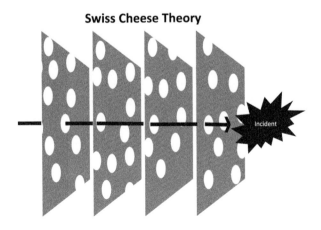

FIGURE 9.1
Swiss cheese model.[2]

is that the cheese moves, the holes in the cheese move and new holes are created. But the most important lesson, perhaps, is that there are holes in the first place. It must be remembered that "low risk" doesn't mean "no risk". In effect, a risk assessment is a map as to where the known holes in the cheese are, the countermeasures that are available to best combat those holes (i.e. to prevent that critical line-up). Scheduled risk review or incident investigation provides an opportunity to add to our knowledge any new holes found and also provides an opportunity to plug them if possible. Thus, the review of a risk assessment is a stock-take on the known holes and an inspection of the cheese for obvious deterioration. All too often risk assessment review is seen simply as changing the date and management signature.

Where this concept can be used to good effect is where the levels of risk are significantly elevated and strict controls are introduced such as a Permit to Work (PTW). In effect, the PTW should be viewed as a slice of cheese with no holes in it at all – a blind or blanking plate (see Figure 9.2).

This documented process is so safety critical that it must eliminate all potential for incident because the realistic outcome would be a fatality. A PTW must not only take into account the controls for the task at hand but must effectively look at neighbouring cheeses and their holes. An example would be if entry is permitted into a sub-ground-level chamber that has been made safe, but a passing vehicle punctures a storage vessel containing a high volume of liquid that flows directly into the hole in which the team are working.

As we have discussed elsewhere in this book, risk assessments are effectively a prediction of a typical incident that could occur despite our best

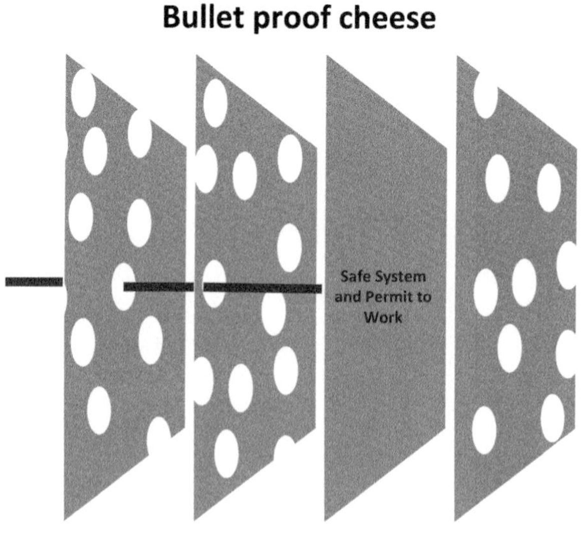

Bullet proof cheese

Safe System and Permit to Work

FIGURE 9.2
Swiss cheese blanking plate.

efforts. Reason's model provides an excellent platform for illustrating this concept and why incidents will still occur despite all reasonable controls being in place.

CASE STUDY

An organisation found that once the reporting escalation approach took hold, they saw a large increase in the reporting of non-injury incidents: most notably in minor electric shocks. The vast majority of these incidents resulted in little or no physical injury, but undeniably the outcome could have been electrocution for every case. These were examples of where the risk assessment, competently performed, could not possibly pick up the potential actions of every single employee. So where the risk assessment for the majority was "suitable and sufficient" and the controls were genuinely adequate, the sheer variety of work being undertaken by such a large and varied workforce meant that it was not always going so well.

Until the new incident reporting system reporting potential level 3s was brought in, there was no way to record these incidents, which must have been going on for some time. Now having this trend information, a targeted approach to electrical safety was undertaken and the situation was brought quickly under control. It could be argued that a lagging data source was being used to prompt investigation into the effectiveness of controls and thus lead action taken to rectify the weaknesses found.

The problem of under-reporting

Under-reporting can be caused by many things:

- pressure to improve accident statistics;
- an unreasonable aim to achieve zero accidents;
- pressure to maintain the "number of days worked since the last lost time accident" statistic;
- the environment is very busy – "we'll get to it later…"
- leaving it to the next shift;
- general complacency about safety by either employees or management;
- a belief that there is no point because we never learn from the investigation findings.

In organisations where claims for compensation may be the norm, reporting will only occur when an injury has been suffered, i.e. grounds for litigation. Even these organisations will not provide the wealth of information that can be gleaned by the adoption of a RFWCI approach to drive reporting and investigation requirements.

We are sorry to say that there is no simple formula to apply to improve incident reporting, but there are some considerations that will encourage it:

- Don't expect overnight change. In most organisations, this is going to take some time to embed. There will have to be a number of key stakeholders in each location competent in the evaluation of the RFWCI approach to assist with tempering some of the reactions to incidents. There is always someone who fearfully wants to record every splinter as a potential fatality erring on the side of caution just in case they might get it wrong.

- You need to have a risk assessment approach that will complement this way of looking at incidents. There is nothing fundamentally wrong with using the HSEs Five Steps to Risk Assessment and their guidance in HSG 245 Investigating Accidents and Incidents, unless of course you want them to blend seamlessly together.

- The timing for a launch of such a system has to be right. In our experience, the organisation must come to terms with a task-based risk-led system first. It may be worth picking an area of the organisation to trial the new approach first. Success may provide leverage to achieve similar results elsewhere.

- To start with, accept everything that is reported and use it to educate. Sending people away will serve only to disenfranchise them, which is the opposite of what you want. Publishing good guidance on the "what and how" is essential here; use plenty of good examples from your organisation's history to illustrate what is appropriate.

Two further myths to discuss. First, having an incident does not necessarily mean that controls need to be changed or added to. It is entirely possible that an investigation concludes that it was a one-off event and that no changes are necessary. Second, despite many people saying that their objective is zero accident, realistically this will never be realised. That drives risk aversion and can encourage unhelpful blame cultures and even force organisations to take action that is beyond what is reasonably practicable. In any case, the easiest way for managers to achieve incident reduction goals is to forget to report; clearly, that would be very unhelpful for any meaningful progress.

Accident triangles: fact or fiction, help or hindrance?

A recognised theory first proposed by Heinrich[3] suggested that incidents can be analysed and a ratio of numbers indicating levels of seriousness could be identified. This theory has been supported by work completed by others down the years such as Bird.[4] This has perpetuated a widely held belief that incidents can be viewed in this way. The hypothesis suggests that if you could reduce the number of non-injury or minor injury incidents, then you would make the chance of suffering a major injury event more remote. We would urge the reader to consult Manuele's[5] evaluation of Heinrich's theory. He makes a very clear case for why Heinrich's theory and the conventional reliance on accident triangles is fundamentally flawed and unreliable for use today. We would add that such a theory does not take into account the different causalities of the incidents within the triangle. It's like suggesting that the numbers of ice creams sold and shark attacks must have correlation since both are at the beach.

Manuele's argument includes evidence that ratios are potentially harmful since they prompt the safety practitioner to focus on trivial incidents rather than concentrate on identifying potential catastrophic risk. The findings of the Baker Report[6] on BP's explosion at the Texas City Oil Refinery in 2005 support Manuele's view. Manuele further suggests that modern-day safety professionals find little evidence that fatal or serious injury incidents are linked to frequent minor injury ones.

We would add to the Manuele and Baker view that in our experience, non-injury events are rarely recorded by organisations and seldom investigated, despite many organisations having blanket policies to report them. Probably because almost any event could be construed as a near miss, there are huge numbers every day in any organisation. That makes the true base of the accident triangle difficult to identify.

Setting goals around incident numbers can also be flawed. If we set a goal for reducing the number of incidents by 10% this year, are we suggesting that we are happy with the other 90% happening? What if the trivial injury events were in the 10% and the major injury events still in the 90%; what would we have really achieved? What if you reach your goal within 6 months, can you then relax for the next 6?

Using incident information in board reports

Having reporting systems that pick up such occurrences demonstrates that even in an intrinsically low risk business like retail, near miss fatalities still occur. Board members need to be appraised of this if they are to lose the blanket belief that all is rosy in the garden because most actual injuries reported are minor.

BOARD REPORT EXAMPLE EXTRACT

"We had 200 minor and 50 moderate injury incidents last year. Of those 35 could reasonably have resulted in fatalities and only 12 of these were reportable to the authorities. Ten of these related to vehicles reversing into the yards and on to the docking bays of stores. This further supports the case for acting on the recommendations elsewhere in this report for implementing new controls for reversing vehicles; our number one risk on our risk register.

Twelve other incidents were electric shocks suffered when working with display cabinets in stores that reasonably foreseeably could have been electrocutions. This supports the training and redesign recommendations given for working with display cabinets; which is also in the top 6 risks on this year's risk register.

Implementing the recommendations for both of these risks will not only potentially reduce the total number of incidents but would also have the following business benefits…"

So, as a profession we believe it is time to challenge the existing view of how incident statistics should be presented in future annual reports. We may decide to place incident numbers in an annex, but we must explain to directors what they mean and how they relate to the significant risks as prioritised on the risk register. What is clear is that merely listing and comparing incident numbers from year to year does not provide a meaningful picture of how we are actually performing in our safety management. It does little to inform our safety management decision-making.

Summary

- A healthy scepticism in incident statistics must be held; faith is better placed in the findings made by competent risk assessors – the true source of leading indicators.
- Organisations must define what incident, accident, and near miss mean for them and be disciplined in the use of the terms.

- An incident reporting process that captures significant near misses is vital and they must be investigated as though the RFWCI had happened.
- Having a focus on what makes the predicted incident more or less likely to occur in the risk assessment enables a direct link with incident investigation.
- A reporting and investigation system based on the hazard scale, actual injury and RFWCI provides a clear action framework and escalation system.
- The Heinrich ratios and Bird triangles are misleading and scientifically disproved theories.
- Setting performance targets for reducing accidents is misleading; systems must provide focus on the types of significant incident, not classification of actual injury.

References

1. Health & Safety Executive (2004), *Investigating Accidents and Incidents, HSG245* (2nd ed). London: HSE. Accessible at: http://www.hse.gov.uk/pubns/books/hsg245.htm.
2. Reason, J. (1990), The contribution of latent human failures to the breakdown of complex systems. *Philosophical Transactions of the Royal Society (London), Series B* 327: 475–484.
3. Heinrich, H.W. (1959), *Industrial Accident Prevention: A Scientific Approach* (4th ed). New York: McGraw-Hill.
4. Davies, John, Ross, Alastair, Wallace, Brendan (2003), *Safety Management: A Qualitative Systems Approach.* CRC Press. p. 45. ISBN 9780415303712.
5. Manuele, F.A. (2003), *On the Practice of Safety*, Chapter 7 (3rd ed). Hoboken, NJ: John Wiley & Sons, Inc.
6. Baker, J. et al. (2007), Report of the B.P U.S. Refineries Independent Safety Review Panel. Accessible at: https://www.scribd.com/document/87023746/Baker-Panel-Report-1to21-BP-Texas.

10

The view from the top

In this chapter, the reporting of risk information up the management chain to the board is reflected upon. The chapter introduces and discusses the value of reporting leading indicators. It will also provide a few ideas of how safety risk can be integrated into the wider risk management considerations of organisations and thereby more fully engage managers. It will also include a discussion of the merits of external accreditation and certification.

Many organisations make the mistake of reporting the wrong things to the wrong level of management. If accountability, responsibility and above all responsiveness are to be achieved, then how risk is presented to different levels of line management must matter. This may well include integrating health & safety risk into the wider risk management system of the organisation. These ideas are explored in this chapter.

In Chapter 6, we discussed how task-based risk assessments and risk registers can inform and direct local management decision-making. Reporting and action get more complex the larger an organisation is or the more sites it has. The first step in obtaining the support of directors is deciding what kind of organisation you are. Are you large, medium or small? Are you situated on one site or many? How many levels of management do you have – few or many? These considerations will dictate how many risk registers you require, who should hold them and, more importantly, how the risk listed on them is reported upwards to the board.

For many small- and medium sized enterprises (SMEs), a single risk register in the organisation may be sufficient for the board and local managers to act on. This would not be adequate for a large business in multiple sites and with complex reporting arrangements. However, splitting the world into SME and large business is too simplistic. Some SMEs have similar complexity in their management reporting to that of larger enterprises, a situation that may require layers of management between the local manager and the board. For example, a chain of shops may have only three or four employees in each unit, but if there are many of them and perhaps of different sizes, then the organisation can be complex.

Nonetheless, the principles to apply in risk reporting to the board and gaining their engagement are similar, regardless of size or complexity. The advice in this chapter is equally relevant to all types of organisational arrangements.

In Chapter 9, we discussed why reporting accident statistics or the number of legally reportable incidents to the board is not as helpful as commonly believed. In this chapter, we consider what should be reported and some ideas of how information can be gathered, evaluated and reported in a way that assists the organisation to further reduce the likelihood of incidents occurring.

The perennial challenge for large and complex organisations is how does the right information get to the right person, at the right time and in a language that they can understand and appropriately respond to? It is a crucially important question. A director does not want to be bothered with trivia that junior managers should have acted on. Equally, he or she does not want a more junior manager to intercept and fail to communicate something that was vital for him or her to know. This point is as much to do with defining job descriptions and responsibilities as it is of effective risk management. This is important to understand since directors do have a degree of liability for the acts and omissions of subordinates.

Effective communication between management levels is crucial

A simple hierarchical model is shown in Figure 10.1. In this model, supervisory management (local managers) are responsible for maintaining an

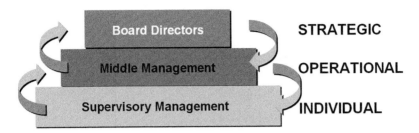

FIGURE 10.1
Management hierarchy model.

intimate knowledge of the task-related risks. They hold direct responsibility for ensuring that the control strategy is implemented effectively for each task. Middle management may have a less intimate knowledge of local task-based risk. This is often due to the fact that they may have a number of localities reporting to them, perhaps with different local conditions. Middle managers are more concerned if the limited organisational resources are being effectively allocated to manage the operation efficiently. For middle management, the consideration is about how all types of risks are managed, not just safety-related ones (e.g. financial, reputation). At the board level, the directors may well wish to keep a watching brief of what is happening operationally, but they are more concerned with forming policy, deciding the mission statement, the annual objectives and overall management strategy.

Reporting task-based risk through this hierarchy must reflect that different levels of management have a different risk focus too. Risk will have to be reconsidered and re-presented to the next management level in a way that makes sense to them. Failing to do so will give rise to the belief or accusation from more senior managers that they are being bothered with trivia that lower managers should be managing for them. It needs a new explanation to ensure that they understand the problem in the right perspective (see inset).

How does the hierarchy in Figure 10.1 work in practice? If a manufacturing company were experiencing an unacceptable number of employees off sick from incidents on and around the production line, simply ensuring that supervisory management imposes the existing controls more strictly may reduce incident numbers. For middle management, they would consider what they must do to assist local supervisory management to reduce the incident rate, but in addition they would also need to consider the effect this was having on production levels, finance, reputation, customer service, etc. The board would consider how the situation was affecting their mission statement, policies and the business objectives set and how these need to be modified to help reduce the likelihood of such accidents occurring. The implications for large and complex organisations are that between each level in the model, risk needs to be reconsidered and other related business risk considerations taken into account to gain a more comprehensive picture. This action will improve responsiveness and help to assure the making of sound judgements and decisions, especially when it is supported by quality-assured risk assessments containing auditable data.

LANGUAGE USE AT DIFFERENT ORGANISATIONAL LEVELS

The modification of language for different levels of the organisation should be done without compromising vision, e.g.:

- *Director*: "Alice, as Operations Director do you realise that if we don't make a safe warehouse a reality, it could seriously affect your profit and loss account. The task-related risk assessment shows that a serious incident is likely with forklift trucks. If we had such an incident, it would mean orders would not be shipped on time, resulting in loss of hard-earned customer confidence and our reputation. In the worst-case scenario, poor publicity could lead to brand damage and loss of market share. We can avoid this by taking a few more low cost actions; some of these ideas may even improve efficiency and reduce costs too..."

- *Middle Manager*: "Bill, as the warehouse manager you realise that if we had a serious incident as a result of this poor forklift truck driving behaviour, it would affect your operation, as you have agreed it is reasonable to predict such an accident. If it happened, we might have to shut that section of the warehouse and take that forklift truck out of action until the enforcement officers had finished their investigation. Managers and employees will be taken off their jobs to be interviewed. We would lose production. It will certainly affect the volume targets you are chasing and we'll all have to work extra hours to make up the shortfall. Can you see the value of putting these controls in place?"

- *Employee*: "Charlie, you do realise if we don't change the way we drive forklift trucks in this warehouse, someone is going to be seriously hurt when one speeds around that corner and over the pedestrian route. If you ran Derek over, would you ever be able to look his family in the eye again? What would you think you'd feel like if you passed them in the street? Now let's have a discussion about how we can avoid such an accident..."

Reporting using risk registers

If the organisation is small, organised in a simple way or has a very narrow portfolio of risk, then one safety risk register will probably be sufficient. In larger more complex organisations with different sites and functions, one safety risk register may be inadequate. The development of local risk registers will prove to be a valuable tool for local management teams. The development of an organisation-level risk register needs to be developed from all reporting units.

In safety risk management systems we have implemented, we have required only the annual reporting of the top six risks from each subservient risk register to the organisation's main safety risk register. This has the following advantages:

- It stops the senior levels of the organisation being swamped with information about all task-based safety risks in the organisation. Many of these are already sufficiently controlled and are the responsibility of supervisory management: that's what they are paid for.
- It has a sufficient number of risks to enable common elements to be identified, e.g. a manual handling task is in every unit's top six risks which may be an indication that we need to review our manual handling controls and safety performance in the control of these risks.
- The bottom three task-based risks help to justify why the top three are more important.

As actions are taken to further control risk, the likelihood of the incident will diminish and risk will be mitigated. When this happens, a risk may drop out of the top six as their likelihood is re-evaluated lower. This will mean that another risk may move up the list and take a position in the top six. In addition, changes in controls or operational circumstances may also change the risk level and risks may be "bumped" down the list by a risk more deserving of attention. Thus, the top six may well change from time to time. That's why risk assessments should be regularly reviewed against potential changes to the task they relate to.

For organisations that have a relatively stable portfolio of risk in well-established operations (e.g. retail), there may come a time when the top six in the risk register stagnate. Nothing more can be done to control them further. When this occurs, the first thing to do is to check whether any risks lower down in the risk register warrant further action to improve control. If that is so, they can be reported forward in addition to the top six risks on the risk register. The top six are still reported as part of governance and assurance to the board. It informs directors that we are still exposed to these risks and confirms that we are doing what is reasonably practicable to control them.

Accountability, responsibility and responsiveness

What is meant by manager responsibility? Published guidance from enforcement agencies is very clear, directors and managers are responsible for health & safety, but what does responsibility look like? Most directors know they are responsible, they may even understand that they can be held accountable, but many do not understand what they should be doing to demonstrate that they are taking health & safety seriously.

The words "responsible" and "accountable" are sometimes used interchangeably. This can be problematic and needs discussion. A good manager will give responsibility to a subordinate with the right competency and resources and therefore the ability to act; however, the delegating manager retains the overall responsibility for making sure that the subordinate acts appropriately on their behalf.

Accountability is the ability to be held to account for the success, failure or otherwise of the objective set for them. If the person given the new responsibility did not have the resources or the competency to achieve the necessary outcome, then he or she could not be held to account for the failure. The accountability would be assigned further up the management chain at the point where the ability to provide resources and ensure competency lies. Failure to achieve where ability was apparent would attract liability and maybe even accusations of negligence. Responsibility is about having (or doing) something and is normally set through job description or project role definition.

To show that they are acting responsibly, managers must be able to demonstrate that they are receiving quality information and are providing assurance to the next level of management that they are responding appropriately to it as far as their resources allow. This underlines the point that managers must present auditable information using the language style of the next level of management. The focus of any reporting and decision-making system must be to present information in a way that enables *responsiveness* from the next management level and ultimately from directors.

So what should be reported to more senior management and directors? Is it as simple as providing accident statistics and information about hazards? As discussed in Chapter 9, accident statistics are clearly trailing or lagging indicators and organisations should be wary of making them the main focus of analysis. The hazards faced quite often cannot be avoided because they are associated with tasks that must be done. Neither hazards nor accident statistics may be helpful in making the board respond. Managers and directors must be able to demonstrate that they are proactive in identifying and controlling significant safety risk.

In the UK, many directors are concerned about their accountability with regard to the Corporate Manslaughter and Corporate Homicide Act 2007. We counsel that if a task-based risk-led management approach for health &

safety is fully adopted, providing that the directors respond to the good data gathered and the management system is audited to demonstrate it is being effectively implemented, then it would be very difficult to charge them with the gross negligence under this Act.

Predictive indicators

Health & safety reports for senior managers and directors are no different from those reported in other disciplines. Organisational leadership require assurance that the safety management system is being effectively applied and that health & safety risk is being successfully mitigated in line with legal requirements and the organisation's risk appetite and tolerance.

What they commonly demand, therefore, are performance measurements so that they can gauge whether the health & safety policy and strategy is being successfully applied, maintained and even improved upon. While this is a need for the health & safety professional to service, there is confusion about how to define a performance indicator and how to interpret its meaning. There is a reliance on lagging indicators by many practitioners and for some the identification of leading indicators has proved to be a difficult concept. It is worth reflecting on this important topic from the perspective of a risk-led, evidence-driven approach to safety.

In Chapter 5, we discussed that lagging indicators or historical data can be used as part of the assessment of the likelihood of an incident. Lagging indicators are the easiest to collect but they can be misleading. While they provide the benefit of hindsight, it is possible to misinterpret their meaning and their relevance to present conditions. All businesses work with a rapidly changing internal and external environment that they must continuously respond to. To thrive, they must be flexible and rapidly respond to ever-changing demands. Externally, pressures like customer demands, legal requirements or competitive positioning drive the need for internal change. Internally, organisations are consistently reassigning or developing new role responsibilities, new reporting structures, new equipment, new ways of working, financial pressures and the volatility of the employment market. The relevance of historical data to the present-day conditions can therefore be undermined by such change. It may be difficult to argue that the organisation is the same today as when the lagging data were generated. The relevance of historical data to the present-day conditions may have a "shelf life". We do not dismiss the usefulness of lagging indicators but merely suggest caution when interpreting their meaning for these reasons.

An exacerbating factor, allowing historical data to be misinterpreted, is the stubborn adherence by some occupational safety and health (OSH) professionals to the tenets of the incident pyramid theories argued by Heinrich or Bird (see Chapter 9). As we suggested, accident triangles have been dis-credited. They ignore that incidents have significantly different causalities throughout the pyramid. These theories infer relationships that, mostly, do not exist. Yet even so, OSH professionals are tempted to use lagging indica-tors to evidence a need to work on reducing minor injury incidents, in the mistaken belief that it will help to make serious incidents more unlikely.

Training helps to develop competence, which in turn reduces the likeli-hood of accidents. Frequently, it is claimed that such things as the number of workers attending a training course or the number of inspections completed are a leading indicator. These are outputs rather than outcomes. Trainees may have attended but failed to understand a key message from their course and returned to the workplace to carry on as they did before. It would be bet-ter to measure any changes in behaviour driven from attending the training or completing the inspection by observation or staff surveys.

Leading indicators are most commonly used to mean two different things by organisations:

- Specifically, to assess the likelihood of a specific incident occurring.
- A means to demonstrate the maintenance and improvement of a positive health & safety culture, in the hope that in doing so acci-dents will be generally less likely to occur.

Which view is favoured depends largely on what the organisation's risk pro-file is and how critical the risk is. If the organisation has a high-risk profile (e.g. construction, manufacturing, agricultural sectors), then more general leading indicators may be sought to indicate a healthy culture and effective management practices. Even in these industrial sectors, there is a great deal of effort and expense required in gathering data, analysing and reporting it. There is little point in doing so if the risks are not critical or are known to be relatively insignificant.

When the failure of a component could lead to a catastrophic failure, then leading indicators are highly relevant (e.g. failure of valves and other compo-nents in high-pressure systems). Techniques like failure mode effect and crit-icality analysis (FMECA) can indicate the potential life of key components in plant or machinery; the leading indicator would therefore be how many of these components were checked and/or replaced before their life expectancy was reached. Other engineering sensors (e.g. oil pressure, temperature) can be engineered into the system to provide leading indication of future failure if action isn't taken. These are commonly used for some systems in cars with warning lights on the dashboard. However, paradoxically, these predictive indicators are based on historical evidence of failure, e.g. product testing results or maintenance logs.

To illustrate changes in safety culture or worker behaviour, organisations often adopt worker surveys to provide leading indication. In doing so, they must ensure that the questions are phrased well to avoid bias and perhaps repeated to avoid vagaries of fluctuating moral. While worker surveys may measure attitudes, they do not necessarily reflect how effectively individual risk controls are being implemented; so it's debatable whether worker surveys provide leading indication of incident numbers or severity. But they can be engineered to provide useful insight into how well risk is understood and the attitude towards applying controls.

Does reporting audit results to organisational leadership provide leading indication?

Audits are snapshots in time that measure existing or recent experiences. In that sense, they uncover historical data and information. There are different types of audit:

- *Legal compliance audit*: Does the organisation comply with the minimum standards required by law?
- *Internal standards audit*: Does the organisation comply with its own internal policies, procedures or standards?
- *Gap audit*: What's missing from what we do that's not covered by legal or organisational requirement?

Gap audits seem to be the closest to providing leading indication, while the other two scrutinise what the organisation has been doing: that's lagging or historical.

The results of audits are not indicators themselves, this is provided by the evaluation of the findings and the extrapolation of their meaning. Certainly, this process can generate actions that may be taken to develop and improve. In this sense, audits can be argued as being very similar to incident investigation where factors are examined in close detail and an analysis is undertaken of the likelihood of reoccurrence.

Near misses, or the observation of unsafe acts or conditions, have been argued as being both lagging and leading indicators. Perhaps it is luck that meant an incident did not cause injury, so for this reason some argue they are lagging indicators. Others say that they are opportunities to identify if controls are not working well, in which case they are leading indicators. This is one of the reasons why we respectfully suggest that dropping the terms leading and lagging indicators may be beneficial. We suggest just collectively referring to them as predictive indicators.

Consideration must also be given to the validation of the prediction. When indicators are balanced against one another, a different picture can emerge. If an improvement is observed like workers changing their behaviour after training coincides with a spike in the numbers of related incidents, it is a clear indication that the wrong things are being measured. Justification for action should be provided by the concert of several complementary indicators that corroborate each other. A blend of information sources provides a more reliable foundation for making indicative judgments, e.g. accident history, inspections, audits and observations.

As a final point on this contentious topic, task-based risk assessments themselves may be regarded as predictive indicators. Boards that receive health & safety reports focused on the presence of hazards or past accidents do not have all the information that they need to reason that the safety management system is being effectively applied. Adopting a risk-led philosophy as recommended in this book changes the focus to the evaluation of the likelihood of accidents occurring. The evaluation of the factors considered for likelihood (see Chapter 5) provides predictive indication. If operational circumstances change and the effectiveness of controls deteriorate, then the likelihood of having an accident will increase. Over time, therefore, the trending of likelihood calculation in a risk assessment arguably provides predictive indication.

Integration with wider risk management

One of the tensions between the health & safety function of an organisation and operational managers is related to perceptions and knowledge of risk in its wider sense. If true risk management is to be encouraged, it is right that task-based risk is placed in the context of the other risks that the organisation is facing. It would be foolish to press for a major refurbishment of the warehouse floor if the organisation is facing tough competition and the priority should be to invest in a new computer booking system to stay competitive. It is no good being the safest failed organisation in the world. Of course, you will have to look at what can be done, e.g. complete patch repairs of the floor and plan when the full refurbishment can take place. This argument is again one of reasonable practicability.

Most would agree that a problem is rarely purely a health & safety one, it also includes implications for the efficiency of the operation, morale of employees, management time, finance, production efficiency, etc. If this is true, then we must also argue that non-health & safety problems, like modifying a machine for greater efficiency, must have a health & safety angle to be considered as part of the final solution package. It is appropriate, therefore, to balance judgements of the different types of risk to find an effective and comprehensive control strategy.

Rather than traditional health & safety committees, the formation of a health & safety risk management co-ordination group(s) to consider task-based safety risk in the light of other operational risks and objectives can be helpful. The membership of this group should predominantly be made up of operational managers from different functions. It is their collective knowledge of present and future business objectives that is important. They will help to form the right context for the content of the report for the board. They will also be instrumental in accurately analysing and evaluating predictive indicators that will inform the report. In this way, they provide the board with assurance that health & safety is integrated into the objectives of the business. These groups help to ensure that safety is not regarded as an insurmountable obstacle to projects, or something that is a separate consideration, but instead is considered at all stages of a project during planning, delivery and implementation.

Further discussion about integrating risk management is beyond the scope of this book, however, we do suggest one simple technique that can be utilised, as shown in Figure 10.2. Organisations can develop their own versions of this impact comparison table. The number of impact ratings will depend on the risk methodology adopted by the organisation; three are demonstrated here to be in keeping with the methodology used throughout this book. The number of columns will depend on how many impact issues the board feel they must consider. Four are shown in Figure 10.2 but others could be added, e.g. marketing or environmental risk. We offer these example levels of impact for illustration, the words and levels would depend on the type and size of organisation, the impact category, etc. There are no rules as to what these boxes should say. Organisations can set these parameters to suit themselves. Remember what we are doing is further refining the prioritisation of what we already have identified as significant risk. They help us with the calculation of risk, which is a label of relative significance.

It would not only be impacts which are considered but the likelihood of the events occurring through failure which will provide value to risk management discussion – discussions about how well we are presently controlling the likelihood of each event occurring, if that is reasonably practicable

Impact Rating	Finance	Reputation	Health & Safety	Continuity
3	Loss >£100k	National press exposure	Death, serious injury >5 days lost	Inability to trade effectively
2	Loss £25k–£100k	Extended local press exposure	Moderate injury, medical attention <5 days lost time	Interference with operations across the business
1	Loss <£25k	Short local press exposure	First aid events	Local interruption to operations

FIGURE 10.2
Balancing impacts.

and whether our risk appetite makes that tolerable or pushes us to do more if we can. An integrated response to risk in this sense means the ability to measure which aspect of the whole problem can be or should be addressed first. It also means that organisations should use similar risk assessment methodology for different risk types, i.e. Task (context): Impact (hazard) x Likelihood of the event = Risk.

Seeking external validation

We cannot leave a discussion about reporting and the view from different levels of management without further considering the merits of external validation certificates (see Chapter 1). Boards will often request such accreditation. What does having this type of certification actually mean?

The first question is to ascertain what the board believes the benefit will be: achieving the standards required and driving consistency in different business units or having the award. If it is a belief that it demonstrates that they are managing health & safety competently, this may not be true. It is possible to have the best documentation and reporting system in the world that measures well against the certification standards, but if the system it promotes misses hazardous tasks or conditions and if it is not in tune with the organisation's structure and operational reality, then it is nothing more than a generation of costly bureaucracy bringing false hope. Typical pitfalls include the following:

- Despite the authors of certificated standards openly stating that they are only meant to be a guide and must be adapted to operational circumstance, organisations don't always feel confident in how far they can interpret the standard and so fully implement it anyway.

- Organisations can over-rely on consultants to tell them what is required even though consultants often don't have sufficient intimate knowledge of the organisation's business arrangements or operational context. Of course, the consultant also has a vested interest in selling more support and creating dependency.

- Another major pitfall with reliance on certification against standards is that it is possible to concentrate more on compliance with the system rather than with what you are trying to achieve – striving for the accolade rather than making things safer in reality. Certification becomes the sole focus and usually results in needless increases in bureaucracy. In turn, this may make things complicated to understand and managers can fail to act, thereby increasing the likelihood of accidents.

- It is sometimes claimed and often assumed that applying these standards will reduce numbers of accidents. Of course, they can't. It is the actions of managers and employees that reduce accidents – bits of paper don't. Neither does a nicely framed certificate, a file of pristine risk assessments on a shelf or a glossy H&S handbook for managers. They are irrelevant if the reality is that employees and managers are doing things differently on the ground compared to what the paperwork says.

Great care is needed when employing external standards, accreditation or certification systems. The reader may surmise that we are not fans of meeting voluntary national or international standards. While we are sceptical, we would recognise that if approached in the right way, they can be beneficial in providing assurance and consistency across the operation. We have so often seen organisations that have generated a nightmare bureaucracy for themselves and are still experiencing similar incident levels. The bureaucracy is often driven by a notion of continuous improvement – leading people to believe that just because a control can be conceived, it must be implemented, whether actually needed or not. It is possible to go too far to the point of becoming unsafe again. One organisation we know of had 27 policies and associated procedures relating to health & safety issues. There was so much information that busy operational managers could not memorise all of that information and it was too much to go through every time an operational decision was necessary. This resulted in the policies and procedures being ignored in favour of "doing what I think is right". The complexity undermined confidence and performance. Thankfully, the guidance to the new ISO 45001 standard makes it clear that continuous improvement includes simplification.

In summary, the key points made in this chapter are as follows:

- The organisation's reports and reporting process must enable managers and the board to be responsive.
- To enable responsiveness, task-related safety risk is better presented with the consideration of other related risk impacts.
- The language used to describe task-related risk should be appropriate to the recipient and their management level.
- Lagging and leading indicators provide important information for reports, but there are common mistakes made in their adoption. It is the analysis of a blend of them that provides more accurate and reliable predictions.
- An organisation that wishes to proactively reduce the likelihood of accidents must identify and report leading indicators and place less emphasis on reactive data.
- External validation through certificated standards should not lead the organisation to needless costly bureaucracy and care must be taken that it does not produce a false impression of reality.

11

Creating desire and managing change

No handbook, basic introduction or not, is complete without a few ideas on how to implement the risk-led approach to health & safety. This chapter provides advice on how to inspire others and encourage desire for the benefits of a risk management approach. The basics of how to plan a change programme to introduce and implement the ideas packed into this book is included in this chapter. Its content is founded on the hard-won experience of the authors acquired during time as consultants and as corporate health & safety managers.

The chapter describes seven simple steps for introducing a risk-led approach into an organisation, reflecting some basic and common project management considerations. This includes planning to celebrate success and maintain things into the future.

A point made several times in this book is that occupational health & safety is just one more facet of accountability that managers should be managing. All too often managers are under the mistaken belief that "the health & safety professional does safety, not me". Of course this mistaken belief is easily discredited by asking, "What authority does the safety professional have over operational managers?" How can someone be responsible for something that they have no decision-making authority over? So if the premise to this chapter is that implementing a new task-based risk management system like the one described in this book involves obtaining management buy-in, then the health & safety professional must develop a plan for achieving that.

Communication is key

Before we outline the steps needed in a plan, consideration must first be given to how the health & safety professional communicates with their peers around the business. Using the technical language of the discipline has limited effectiveness. Finding a way of communicating effectively with operational and financially orientated managers is vital to success.

You will recall a version of Figure 11.1 in Chapter 7 (Figure 7.3). It is repeated here to demonstrate that it can prompt the health & safety professional to communicate in business language that can be understood by colleagues. We argued in Chapter 7 that minimum standards are set by local legal requirements; equally, for international businesses, these minimum standards could be imposed by head office and may be over and above local legal requirements. Minimum standards are important; they set the mark that no unit may fall below. That does not stop an additional and higher level of internal standard being set to stretch units and encourage them to improve their health & safety performance. There is no reason why these lines cannot be drawn at different places for different units depending on the stage of their safety management system maturity.

As argued in Chapter 2 and further developed in Chapter 7, reasonable practicability can be explained as a proportionate response to risk or deciding when things are safe enough. Where this line is drawn on the sliding scale will depend on many factors, e.g. competition, brand, public perception and organisational finances. Note that these considerations are not purely about health & safety. In the main, they are business considerations. Where this line is drawn on the sliding scale of safety will depend on the business discussions of what level of risk is acceptable and what degree of residual risk can be tolerated. This is a business conversation, not one purely about health & safety morality but what is safe enough.

In moving the internal standards line even further to the right on the scale, to the point of excellence, will require the investment of additional resources,

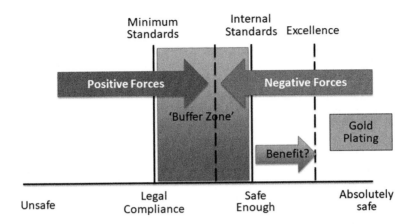

FIGURE 11.1
Safety as a sliding scale.

perhaps more than those applied by competitors. This could be a distinction that sets the organisation apart from its rivals: a unique selling proposition. To purposely go beyond what is regarded as the norm, to argue standards of control others may regard as "gold plating" could bring additional benefits. It is a business and not a health & safety decision. The organisation is already compliant with the law, but it wishes to not only go further than the law requires but also further still to gain some kind of business advantage. The business needs to decide whether it wishes to be a compliant business, one that is comparable to the best or if it truly wishes to be seen as world class. It's a business decision.

As described in Chapter 7, most units at any given time will be somewhere in between minimum and internal standards (as shown by the dotted line in Figure 11.1). Health & safety audits will determine where and action plans will be made to prompt improvement. Unfortunately, audits, inspections and observations often lead to action plans centred on improving health & safety arrangements, procedure and controls. This perpetuates the general misconception that health & safety is somehow extra and separate from the daily operations.

What many fail to fully understand is that the positive and negative forces that push the dotted line back and forth are often not directly safety orientated but often have a business reason.

Negative forces may include late orders, staff turnover and subsequent loss of competence, new management, restructures, changes in business policy affecting culture and well-being, deterioration in staff relations, cash flow problems, new equipment, old equipment breaking down, contractors overrunning, changes in procurement arrangements, poor quality components from the supply chain, etc. The list is almost endless. Positive forces may include new training programmes, better equipment, positive changes in the environment, improvements in communication between managers and workers, improvements in space or lighting, reorganised traffic routes, better maintenance regimes, etc. Again, the list is almost endless. Note in both cases, however, these are changes to the business and not strictly to the tasks or the hazards workers are exposed to. They are factors that can affect the likelihood of incidents occurring. All of these factors are business considerations. It underlines how important it is that the health & safety professionals develop their business language, skill and behaviours in parallel to the development of their technical competence. Without knowing how the business is structured, the different roles of its components, how it operates or the business challenges it faces, the health & safety professional's influence will always be limited.

Seven steps

Having set the scene that the health & safety professional needs to adopt and use business language to influence change, we propose that there are seven steps to implementing a risk-led and evidence-driven safety management system. Of course, these different steps may overlap or even run concurrently, but for ease of explanation they are described separately and distinctly from one another.

Step 1: research, credibility and stakeholder recognition

The health & safety professionals need to do their homework on the organisation. They must understand the roles of different departments and functions and also key management roles. Things change more readily and are more likely to embed if full collaboration and agreement are made with other business functions. Technical expertise alone is insufficient for the health & safety professionals, they must be competent in business skills too. Knowledge of other business functions is vital if language is to be modified into a form that will be understood by different managers if communication is to be effective. This will include the identification of stakeholders: those managers in the organisation who may be able to help or hinder. The health & safety professionals must sell themselves to these stakeholders as good listeners, good educators and pragmatic problem-solvers. Better still if they can add enthusiasm for business aims and realistic suggestions of how the proposed changes will directly benefit the organisation in ways other than improving safety performance.

Personal credibility is essential if the health & safety professional is to get his or her message across and be heard. Without it, managers are going to be unresponsive at worst and obstructive at best. No change programme can be embarked upon without credibility. Stakeholders have to believe that their health & safety professional is one who

- understands the organisation and its operation with some intimacy, particularly the operational problems he or she faces. This includes getting to know the roles, responsibilities and accountabilities of stakeholders and the operational problems they individually face. Generating empathy is vital in this part of the process;

- will work with others to find the most cost-effective pragmatic solutions to problems;

- will back managers in arguments of reasonable practicability, i.e. on occasions will help to formulate the arguments for not putting in place an additional control when reasonably practical solutions are already in place and the likelihood of accidents occurring is already low. Just because it is possible does not mean that it is necessary; and

- recommends solutions less in terms of compliance and more in terms of business benefits.

This initial work would also include an understanding of the organisation's risk appetite and tolerance overall and of the individual stakeholders involved too. The organisation, or different parts of it, may be intolerant and fearful of health & safety risk, or it may hold a cavalier attitude, or be somewhere in between. Information from this analysis would influence how the changes need to be introduced, i.e. demonstrating where there is needless fear or exposing the reality of being too cavalier.

An important consideration at this opening stage is to identify what is already good in the organisation. Credibility is lost if change is recommended needlessly, but it is gained if praise or acknowledgement can be given to areas that already meet the new standards that the change programme will set out to achieve.

Before the health & safety professional goes any further, it is vital that a short, simple, and clear explanation of the vision must be developed. What is it that you are setting out to achieve? This will be different from one organisation to the next but without being able to capture the vision in, say, 100–150 words will only hamper progress. It needs to include statements that hook managers into wanting what you want. The vision statement must include an explanation of where we are, what needs to change and what the benefit of that change will be (see example in the inset).

EXAMPLE OF A HEALTH & SAFETY VISION

At the moment, we have far too many risk assessments and we are not even certain we have the right ones. This is preventing us from prioritising action and effectively targeting our resources. We need to implement a method to systematically identify and assess risk, focus on what is most significant and apply our focus on reducing the likelihood of incidents occurring. This will reduce the workload for all managers and employees by reducing the number of risk assessments and eradicating needless procedure. This work will better enable us to adapt quickly to market led changes, helping us to stay both competitive and safe. Not only will we save operating costs, but these changes will also reduce our employee and public liability insurance premiums in the medium term.

Description of the vision may need to be altered for different audiences, but the underlying message must be the same.

Step 2: identifying obstacles and planning

Military doctrine suggests that successful commanders always seek to know the enemy's tactics and the effectiveness of their fighting equipment compared to their own well before the battle plans are drawn up. Predicting how the enemies might react to given situations and what they might use to counter attack greatly assists with the overall battle plan and with contingency planning. While describing stakeholders' as the enemy may be a step too far, in principle, the health & safety professional needs to take a similar approach when considering how stakeholders might react to recommended change. What might they do, what can be done to prevent them manoeuvring in this way and what should be done to counter any action they may take?

One way of diffusing the effect of obstacles is to identify them in advance and then tell the stakeholders before they tell you. Better still if the health & safety professional can offer solutions to these problems in advance. Even better still if those stakeholders who are responsible for raising objections or for managing the problem areas are consulted in advance so that they can help identify and be a positive part of the solution. Involvement makes it difficult for them to disagree with the solution and places them in a position where they will be more likely to assist if things do not go according to the plan. Thus, involving stakeholders in the finer detail of the plan generates the feeling of "we did this". In fact, publically praising stakeholders makes them and others more likely to lend support in the future, so don't be afraid of giving others credit (see step 6).

Above all else, any documentation or communication on the project should be simple and clear to the user or recipient. Complicated, jargon-led and technical documentation delivered in pages and pages of documentation will produce confusion and will consequently detract from the project objectives and ultimately dilute success.

Before embarking on generating desire in others, there is yet more work to be done. The health & safety professionals need to know what they want to achieve and the quantifiable benefits it will bring thoroughly and with clarity, e.g. a reduction in administrative time, improvements to process efficiency and financial savings. This should include anticipating the foreseeable obstacles to the change and rehearsing the arguments of persuasion, i.e. addressing the typical management challenges of "it's too difficult", "what we have now is good enough", or "I'm too busy". One of the key aspects to illustrate when communicating with stakeholders is to point out any gaps and weaknesses in the present system and how the changes will significantly improve things.

One of the most common reasons why health & safety initiatives fail at the concept and planning stage is that the organisational leadership are unable to envisage how the aspiration will be achieved. Too often health & safety professionals are tempted to offer a vision of utopia. While boards understand the vision, they cannot picture how to achieve it. A greater degree of realism must be adopted. Successful health & safety professionals have learnt to use patience and developed an ability to incrementally build success. They ask themselves: in year one what the simple things we can achieve? How do we build on this in year two? What in year three? And so on. In this way, the plan will build momentum and be seen to achieve greater and greater levels of success. What is more, when presented in annual increments, directors are better able to envisage what needs to be done next and be satisfied that it can be realistically achieved.

Step 3: creating desire and managing expectations

A universal truth is that many people resist change, probably because they perceive that it will bring a fresh workload for them to do. Without creating a desire for the change, implementing a new safety management system will be much more difficult. The health & safety professional does need to exercise some restraint though; stakeholder expectations need to be managed too. If they are to be successful in creating desire, stakeholders may demand more than is possible or necessary and end up disappointed. It is perhaps better to sell a little short and then deliver in excess of expectations in the end. Using words like "refresh" or "enhance" and phrases like "seeking further benefit" rather than "change" is helpful with stakeholders; such expressions are more difficult to argue against. What stakeholder wouldn't want to enhance systems to realise greater efficiencies or benefit?

At this stage, leadership skill is exercised to try to create a perception of value that will be desired by stakeholders who in turn wish to work to achieve it. What can be helpful is if the health & safety professionals' preparatory work during step one enables them to link the safety risk benefits to better control of other risk considerations too – looking for the win-win situations. If the changes can help the stakeholders to achieve progress on some of their other problems, then they will be more open with their support. They will be more likely to accept the short-term sacrifices that will be needed to realise the long-term gains.

The health & safety professionals must consider which of the stakeholders they have identified have significant influence in the organisation. A respected stakeholder who normally champions safety would be a good ally as usual, but perhaps turning the one who has a reputation as a safety sceptic or as a difficult character but nonetheless agrees to champion the changes would provide stronger cultural influence. The beneficiaries of the changes such as employees, customers and regular contractors should be involved too. If they also desire the benefits that will be realised, then this will provide additional pressure on the organisation to accept that change as necessary.

We can't stress enough how the old adage holds so true – prior planning and preparation prevents poor performance. Implementing change in philosophy, thinking and information will only generate challenges. The health & safety professional must be well prepared in order to meet them with confidence.

Step 4: educate

Once the health & safety professional is sure of what changes are needed, before the plan is implemented and during its implementation too – educate, educate, educate. This doesn't mean just writing and delivering training packages, but by using any means at their disposal the health & safety professionals must also communicate the vision and benefits sought. Training packages can deliver the basic knowledge needed to understand why change is necessary, what will be different and what benefits will be hoped for, but even if the delegates remember what they have been taught in the classroom, they will still generate more questions afterwards. Stakeholder enquiries made by telephone, e-mail or in meetings must be met not only with an answer but also why the answer is correct in the light of the new task-based risk management system. Questions provide further opportunities to promote the changes and educate why they are necessary.

There are aspects in this book that the organisation's stakeholders must be trained in if they are to understand, utilise and operate the task-related risk management system effectively and to advantage. The detail that is included in the training and the language used should be tailored to different audiences. Employees will want to be reassured that risk has been identified and good control implemented. Supervisors will need to understand how to complete task-based risk assessments and decide upon the risk controls necessary. Middle managers will need to know how task-based safety risk will be reported to them, how to place it into operational context by considering other risks to obtain a full picture and how to identify predictive indicators and make recommendations to the board if necessary. A training course for managers should include the following:

- Why is the development and maintenance of competency key to the whole approach?
- How do you know if your judgement is reasonable?
- How do we judge if a risk is significant or not?
- How do we decide that we have controlled the likelihood of accidents satisfactorily? What does safe enough look like?

- What benefit do stakeholders get from being able to define different risk levels?
- How should risk be reported and how do we ensure that it is appropriately responded to?
- What are the benefits we can enjoy from managing safety this way?

The board needs training too. They need to know how to seek assurance that the system is working effectively and how to respond to recommendations, including how to say "no".

Step 5: implement and engage stakeholders

The plan should not be too ambitious too early. It is generally better to implement a small trial or a phased roll-out system, especially in bigger organisations. Concentrating resources to gain some early successes will provide evidence that the new system works and the promised benefits can be enjoyed. The health & safety professional's project plan must include time to spend in establishing the new system in the departments or areas after training people and implementing the system. Although the managers in the starting areas should understand the system and be using it, they will still need further support while they gain confidence in the use of the system (see step 7).

Carling and Heller[1] describe the work of world-renowned sports coach Frank Dick. He describes how a relationship develops between an athlete and a coach:

- "Initially the athlete does not have the knowledge and is dependent on the guidance and education of the coach. The athlete places their full confidence in the coach and may mistakenly believe that the coach can bring them instant results. The coach must rely on their credibility as an educator and manage expectations.
- Next, the athlete starts to realise the extent of what needs to be done and that it will be hard work. This is a danger period since results are far from instantaneous and the athlete realises that hard work is needed from them. The athlete is still dependent on the coach and may be frustrated by the degree of that dependency. Without the coach providing encouragement and identifying even the smallest of successes the relationship can break down in this period. The coach must create desire and manage expectations, identify problems in advance and plan how to overcome them with the athlete.
- Eventually the athlete will grow in skill and become independent. The athlete develops a personal understanding of what needs to be done and what they have to do to achieve it. The athlete becomes confident in their new abilities and becomes more self-reliant. The

role of the coach changes once more to providing guidance to the athlete in how they should structure their work programme.

- The athlete matures and realises that although they can find the answers to most of their problems themselves, they still need the coach to provide an independent view of their performance. The coach may also be learning from the athlete at this mature stage of their relationship. The relationship has moved to become inter-dependent."

There are obvious parallels for the health & safety professional implementing a new project. Read those four bullet points again, but substitute the word "coach" with health & safety professional and the word "athlete" with manager stakeholder; it is just as relevant. Some may be fearful that to follow this approach could engineer the health & safety professional out of a job: "If they did it themselves what would I do then?" This is a common challenge but surely there is a great value to the employer to maintain the employ of those who have the skill to develop skills in others. In any case, there will always be a need for a professional for such things as advise on how to interpret health & safety law, training newcomers, reviewing documentation and for evaluating audit results to ensure that the organisation is not deteriorating in effectiveness. The role of policeman and advisor will never diminish.

One way of engaging stakeholders is to change the style and perhaps content of the health & safety policy. Many policies contain such detail that it is difficult to ascertain whether they are a policy, a procedure or a guidance handbook. Writing a policy in terms of risk management not only future proofs it (as risks can change) but also defines responsibility better in terms of a task-based risk management system. An example extract can be seen in the inset.

EXAMPLE OF RISK-FOCUSED HEALTH & SAFETY POLICY ENTRY

Supervisor is responsible for ensuring the following:

a. The department is using the latest versions of the risk management documentation.

b. Significant risks associated with working activity are identified and assessed.

c. Risk assessments are quality assured, signed by the senior manager and reviewed when prompted by audit.

d. Employees are consulted with during the completion and review of risk assessments and when designing additional local risk control strategies.

e. Risk control strategies are effectively implemented by the unit management team.

f. Any new hazards or tasks are identified and reported to the Health & Safety Department, which will then consider their inclusion in the risk assessment system or advise on how existing local documentation can be amended.

g. Local resources are appropriately targeted to control and monitor safety & occupational health risks according to local priority.

h. Appropriate communication arrangements are made in the department to ensure that health & safety risk information is passed to employees.

i. Where appropriate, safety risk considerations and action requirements are included in local business and operational planning (e.g. planning for seasonal trade or events).

j. Significant incidents are investigated, recorded and associated risk assessments are reviewed.

k. Ensure that contractors are managed effectively while in the unit, with particular attention to signing in and out arrangements and the new risks they may introduce into the working environment.

l. Equipment and building infrastructure are kept in a good and safe standard of repair and are properly maintained.

m. Local emergency equipment is kept in working order and emergency exits are kept free from obstruction.

Our final piece of advice in this step is to work on the macro first and then on the micro (see Table 11.1 for examples). There may be more than one micro to a macro, but the table serves to illustrate the differences and how the system can be embedded.

Step 6: celebrate success

Although celebrating success has been mentioned a couple of times so far, it is of such importance that we believe that it is a step in its own right. Celebrating the success of others in the organisation who have accepted and implemented the new system and approach will help because

- it encourages others to aim for the same benefit and accolade;
- it diffuses the arguments that this can't be done, is too much effort or does not produce the benefits claimed;
- making others look good encourages them to further engage with you in the future, especially when embarking on a new project of their own which reduces the frequency of demand for last-minute reactionary responses as an afterthought in the future; and

TABLE 11.1

Macro, micro & change

Macro	Micro	Change
Training the knowledge needed to operate the task-related risk management system effectively.	Ensure that each trainee implements the task-related risk management system locally and uses language correctly.	Use of safety and risk language more accurately. This supports a helpful change in dialogue through enabling more precise communication.
Complete the task list with the aid of local employees and managers.	Ensure that significant tasks with insignificant risk are listed and annotated with no significant findings. Ensure that a team is assembled to periodically review the content of the list.	Better understanding of which tasks we should focus our attention on, providing greater clarity for local discussion and focus on the most significant tasks with significant risk.
Complete task-based risk assessments.	Ensure that the findings of the risk assessments are incorporated into procedures and their training. This may include alteration of the list of competencies for each job role. Task-based risk assessment and procedure documents need to be cross-referenced.	Understanding the rating of a task-related risk will change the approach of supervisors and employees to the tasks demonstrated as being most significant, thus supporting consistent behavioural changes in the way employees work.
Complete local and organisational risk registers.	Ensure that local managers have risk management plans and are progressing with their implementation.	All stakeholders consider risk as part of their discussions and planning for operational changes in the workplace.

- it promotes your credibility in the organisation. If you have credibility, then so does the system and approach; the relationship is indivisible in most stakeholders' minds.

Look for any way possible to praise and celebrate. Use opportunities in meetings, on e-mails and in newsletters and the like. Make a point of making a visit to thank everyone for their hard work and reiterate the benefits they and the organisation will now enjoy. All such communication reinforces the benefit and helps maintain the momentum for positive cultural change. Let's face it, it's the right thing to do when others have put so much effort into assisting you to achieve your goals.

Step 7: future proofing

Arguably, the danger point in any project is not developing sufficient desire for change or even during the implementation of the project plan; it's at the end when everything has been implemented. Without the means to check that controls are in place properly and continue to reduce the likelihood of incidents occurring, over time things will certainly degrade. Part of the health & safety professional's plan must be how to safeguard the good that has been done and maintain the business benefits that have been realised. In safety terms, this means designing an audit not only to check that documentation is being completed but also to evaluate how well managers and employees are using and implementing the system.

The audit findings and information on leading indicators gleaned from the likelihood information on the risk assessments must find their way onto the agenda of manager's meetings if safety is to be fully integrated to be a part of operational thinking. Managers need affirmation and assurance that risk is being controlled as far as is reasonably practicable. They also need to be prompted if further action is to be taken to either control risk or to improve the system designed to manage it.

Even health & safety committees need to change their agenda to focus on the review and discussion of task-based risk assessments, the effectiveness of local risk control strategies and the overall management of the local risk profile. Consideration must be given to what happens to the agreements and action recommendations that arise from committees or management meetings. Where does the information go next? How do you make sure that it is responded to? Good reporting structures that involve the appropriate stakeholders for the issue being discussed are essential. How this is done, what the reporting structure should be or who has accountability and who needs assurance on what will depend on the size of the organisation and the task-based risk it is exposed to. That's where you come in...

Summary

If you wish to implement the ideas contained in this book (and we sincerely hope that you have found some value in its advice), then you need a plan to enable you to move it all forward. These seven steps provide a framework for doing just that and their implementation has provided us with many successes. Remember though, you will be even more effective if you learn how your organisation functions and if you employ business language in your conversations, reports and business cases.

Here's to your success! Thanks for reading – and good luck.

Reference

1. Carling, W. & Heller, R. (1995), *The Way to Win: Strategies for Success in Business and Sport*. London: Little, Brown & Co.

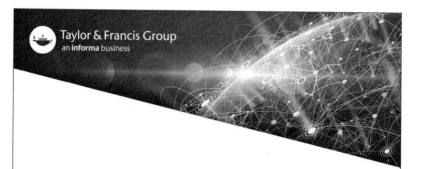